BYPASSES

BYPASSES

A Simple Approach to Complexity

Z. A. MELZAK
University of British Columbia

A Wiley-Interscience Publication
JOHN WILEY & SONS
New York   Chichester   Brisbane   Toronto   Singapore

Library of Congress Cataloging in Publication Data:

Melzak, Z. A., 1926–
    Bypasses: a simple approach to complexity.

    "A Wiley-Interscience publication."
    Includes bibliographical references and index.
    1. Problem solving.   2. Methodology.   3. Creative
thinking.   I. Title.

QA63.M44   1983   153.4'3   82–17364
ISBN 0–471–86854–X

Printed in the United States of America

10 9 8 7 6 5 4 3 2 1

To the gentle memory of my friend
Ian MacDonald

Da dich das geflügelte Entzücken
Über manchen frühen Abgrund trug,
Baue jetzt der unerhörten Brücken
Kühn berechenbaren Bug

Irrlichter
(Rainer Maria Rilke)

(As once the early winged wonder
With you across abysses ran,
So join now those that are asunder
By a magic calculable span)

Preface

This preface is an integral part of the book; the purpose of the preface is to help the potential reader decide whether to continue or to stop. Four elementary examples are given; with these as a basis, the subject matter of the book is briefly set out.

Suppose that a partly damaged Latin text or inscription contains a statement to the effect that XIX times LXXVIII equals M?D??X?I? where the question marks stand for illegible symbols. How would we restore it? The first act would be to translate the Roman numerals into ours, getting the claim that 19·78 equals something. Then the multiplication 19·78 = 1482 would be executed. Finally, the number 1482 would be translated back into the Roman numerals as MCDLXXXII.

Next, suppose that we want a simple sketch of a proof of the following geometrical proposition: If two points lie on a circular cylinder, then the shortest curve joining the two points on the cylinder surface is an arc of a circular helix. Start by unrolling the cylinder onto a plane; since the unrolling preserves length, the shortest joining curve unrolls onto a straight segment. The proof is completed by showing that the reverse operation, that of rolling the plane back onto the cylinder, transforms the straight segment into a helical arc.

Again, let us consider that dream of Sir Francis Bacon in his *New Atlantis* where one of the sages of the House of Solomon says that they know how to "convey sounds in trunks and pipes, in strange lines and distances." Suppose that we wish to outline briefly how Bacon's dream was implemented by Alexander Graham Bell. The principle of telephony might be simply sketched as follows. First, a microphone device changes the acoustic pattern of voice into an electrical pattern of fluctuating current. Then this is transmitted by cable to the destination, and there another device retransforms the electrical pattern back into voice.

Finally, to lend a common touch, we quote a slightly edited version of a conversation overheard in a New Jersey bar: "Those . . .'s in Las Vegas are no fools; they know damn well that a guy will think twice before betting a five, ten, or twenty dollar spot, so what do they do? They make you change your money into plastic chips, now, this ain't real folding money and so you bet like it was nothing. Then you change your chips back into money, if there is anything left to change."

These four preliminary examples have a strong common feature: Each one concerns a moderately complex situation which is effectively simplified by being broken up into three parts, and what is done in the last part is the inverse of what is done in the first. It might be said that the transformation of the first part takes us from one level (Roman numerals, the curved surface of the cylinder, acoustic domain, betting real folding money) to another level (our numerals, the plane, electrical domain, betting chips). The second level is better adapted to the purpose: Multiplication is easier in our decimals than in the Roman system, minimal distances are simpler in the plane than on a curved surface, current in a cable travels better and is amplified more easily than sound in air, and people can be induced to part with plastic tokens that stand for money more readily than with money itself. Then, after the passage to the second and more convenient level, the problem is solved or the task is accomplished there. Finally comes the return to the start, that is, to the original level. Borrowing from standard mathematical terminology, we say that the harder task at the original level is conjugate to the simpler task at the second, and more suitable, level. There is a sort of interlevel round trip: from the first level to the second one, working there, and then returning. We call such a round trip a bypass. This explains the title of the present book, *Bypasses*, and the terminology "The Conjugacy or Bypass Principle"; it also outlines something of the subject matter. To do any justice to explaining the ambitious subtitle, *A Simple Approach to Complexity*, would be a harder and a longer job; it is one of the aims of this book to attempt just that.

By now there should be no difficulty in understanding and accepting the claim that a simple symbolic expression of the conjugacy principle is $W = STS^{-1}$. Here W stands for the harder task at the original level, T for the easier or simpler task at the second level, S for the passage from the original level to the other one, and S^{-1} (read "S inverse") for the return; perhaps the greatest difficulty arises from the equality sign. As might be gathered from that mildly mathematical formulation, the conjugacy principle is rooted in mathematics though by no means restricted to it. In particular, the first example suggests a connection with coding, translating, and language; the third one, with communications and transport nets; and the fourth one hints at some applications in the direction of social sciences.

Three possibilities among serious and extensive uses of conjugacy may be emphasized. The first one is as a unifying principle that might help to counteract the present divisive tendency toward extreme technical specialization. It is surely something important, in mathematics and the sciences, as well as in other disciplines, when a formal connection is produced between things which seemed to be quite different. From this may come the further benefit of using conjugacy as a unifying teaching device. Fields such as mathematics and engineering have grown so hugely in the last generation or two that it is simply impossible even to introduce the student to all accumulated knowledge. This puts an especial premium on a principle such as conjugacy, which brings together under one heading many apparently different subjects.

The second large use of conjugacy concerns research tools and is based on the observation that whenever the mode of calculation or of problem formulating and solving shows field-to-field similarities, there arises the possibility of shared techniques. Now, a technique that is quite standard and ordinary in one field may be transferred into another field and lead there to real advancement. In brief, what we want now is a technical device for exploiting similarities and analogies. It is not surprising that conjugacy may help here too. First, it will be recalled that the theory of similarity and proportion was developed by Thales to enable him to measure inaccessible quantities, such as the heights of Egyptian pyramids. Next, "proportion" in Greek is αναλογια, and in using analogy we also strive to reduce something inaccessible to terms that are themselves accessible. In our first technical example in Chapter 4, we shall argue that proportions, and hence also analogies, are expressible as bypasses. Finally, and as a parenthetical remark, we spoke of transfer of techniques; here it may be recalled that "metaphor" means "transfer."

Perhaps the most important, but certainly also the most tentative, large use of conjugacy is the third one: It might be used to facilitate invention and discovery. It is conceded that many inventions and discoveries in the past were made by accident, and it may be that a guarded and skillful use of bypass could eliminate some of the element of chance. This unlikely sounding contention is perhaps reinforced by the previous observation that analogy itself can be formulated in bypass terms. In further support, there is the methodological role of the conjugacy principle: It is a general complexity-reduction device. Alternatively, it is a method of building up complex process paths out of simple unit steps. These remarks go perhaps some way toward explaining the book's subtitle *A Simple Approach to Complexity*.

It is a pleasure to record indebtedness to many individuals and institutions. My work has been supported by the National Science and Engineering Research Council of Canada (1979–1982). A leave of absence has been granted by my university (1980–1981) and a Senior Killam Fellowship by the U.B.C. Killam Foundation. First part of that leave was spent with my former employer, the Bell Laboratories (Murray Hill, N.J.), and the second part at the Mathematical Institute and the Trinity College, Oxford University. I have had the benefit of help, suggestions, and criticism from: G. Anderson, G. Criscuolo, A. C. Crombie, H. W. Dommel, M. Fagen, J. S. Forsyth, E. N. Gilbert, R. Gnanadesikan, J. Goldman, K. Gottfried, J. M. Hammersley, R. Harré, D. Horgan, M. J. Inwood, A. Jones, F. W. Kaempffer, R. Klibansky, W. Magnus, R. Melzack, S. A. Melzak, W. Messenger, J. D. North, H. O. Pollak, G. Schindler, M. Schulzer, B. Shube, L. G. de Sobrino, J. L. Spouge, A. B. Tayler, H. Weinitschke, and R. H. Wright. My thanks go to all of them, together with a strong form of the usual acknowledgment: The good things in my book owe much to them; the bad ones are my sole responsibility.

Z. A. Melzak

British Columbia
Vancouver, October 1982

Contents

PART TWO.   EXAMPLES IN MATHEMATICS

PART THREE.   EXAMPLES IN TECHNOLOGY AND THE NATURAL SCIENCES

PART FOUR.   SPECULATIONS AND CONCLUSIONS

Gilgamesh and *The Iliad*, Omens and Oracles; Rhetorical Bypass to Induce Emotion; Certain Figures of Rhetoric; A Hypothesis on Metaphors and Similes; Allegory, Metaphors, and Analogies; Metaphor and Proportion as Ways of Handling Inaccessibility; Segmentation and Ambiguity; Bypass and Parsing; A Comparison of Four Occurrences of Conjugacy; Language as Bypass; An Example from Piaget; Language Level and Basic Level; Late Onset of Speech in Future Mathematicians; A Model with Two Interior Levels; A Hypothesis on Consciousness.

Resumé of Conjugacy; Four Important Groups of Bypasses: The Individual–Collective, the Continuous–Discrete–Continuous, the Square-Linear, and the Involutory; Certain Purposes of Bypasses: Equivalence, Transport, and Reflexness; Equivalence Bypass as a Reconciliation Paradigm; Some Remarks on Transport; A Discussion of Reflexness; Hawthorne Effect; Reflex Bypass and Infinities; Secpulations on Human Evolution.

BYPASSES

PART ONE

INTRODUCTION

Chapter One

General Introduction

The principle which is the subject of this book will be called by any one of the three names: the conjugacy principle (somewhat formally and technically), the bypass principle (informally and descriptively), or just the bypass (briefly and loosely). It arose from the author's work started some time ago as an informal though serious attempt to isolate and describe some of the working principles which are in daily mathematical use. However, its range was soon found to extend beyond merely mathematics, first to communications and power technology, then to physics and other natural sciences, and finally beyond science and technology. It is hoped that the whole subject will interest various categories of readers. For the immediate introductory purposes, this variety means just two types: those with less mathematics and those with more. This division explains in part the intended role of the first two chapters: The present first chapter is the general introduction, and the second one is the special mathematical introduction. But hopefully even the mathematicians will start with this chapter and even the nonmathematicians will (try to) continue with the next one. The introductory part ends up with Chapter 3, which sketches the aims of this book, outlines its plan, gives its approximate coverage, and contains some preliminary comments and qualifying notes.

At this stage a definition or at least a description, however rough and provisional, ought to be given and so the following is offered:

(C_0) the bypass principle is a way of dealing with complexity or with difficulty
by means of a bypass which promotes a transport or a passage or the

3

solution of a problem in a three-stage reduction process whose first and last stages are each other's inverses.

This looks circular but the circularity is only apparent: Although the word "bypass" occurs twice, its first occurrence is in our mildly technical sense and the second occurrence is in the everyday sense as a detour or going round. The three-stage reduction process is meant in the simple and standard way and refers to the complexity or difficulty: Reduction means here breaking down one complicated thing into a sequence of three simpler ones. Of these, the first and the last one together will be called the outer part of the bypass, and the second one by itself will be called the middle part of the bypass. According to (C_0), the first and the last stage are inverses. Therefore, it follows that in the absence of the middle part the outer part cancels out; its net effect is nil. This cancellative structure of the outer part will lead to two simple and natural operations: Two bypasses can be combined in length by simple juxtaposition; a single bypass can be extended in depth by substituting a new bypass for the middle part of the old one. A third operation, that of inverting a bypass, will also be introduced. These three operations lead to a rich substitutive and conjunctive structure.

The formulation (C_0) is somewhat rough and vague, a trifle lengthy, and carries a certain derogatory contamination. The latter is accidental: Though bypass may be associated with deviousness and dodging, with us "to bypass" is not "to omit" or "to skip" so much as "to solve" or "to overcome." The examples that will be offered shortly suggest an alternative formulation, even cruder than (C_0) but neither facetious nor entirely trivial, certainly shorter and perhaps more picturesque:

(C_1) the easiest way from one vertex of a rectangle to a neighboring vertex may be round the other three sides.

This describes the conjugacy principle by expanding in graphical terms its single-word equivalent: bypass. There is another restatement, in similar terms but emphasizing purpose rather than geometry and supplying certain details by a moderate use of symbolism; it is

(C_2) a way W of getting past a wall is to dig down a shaft (call this S), then to tunnel under the wall (call this T), and finally to dig up the reverse shaft on the other side (call this S^{-1}, i.e., S inverse); in symbols $W = STS^{-1}$: W is conjugate to T under S.

The clumsy metaphor of this ostentatious statement has its obvious defects too but, as will perhaps be conceded, it has two considerable merits. First, it helps to isolate the common structural factor in the bypass examples to be given, and it produces simple expressions for the three basic operations on bypasses. Second, it suggests right at the outset that the bypass process occurs between certain two different levels. The matter of levels will occupy us at a considerable length in the sequel.

The figure of digging and tunneling used in (C_2) is somewhat arbitrary. S could also stand for climbing up a ladder to the top of the wall; T would then be: bringing the ladder up and letting it down on the other side; and so on. An illustration can also be given without ascent or descent; it is, so to speak, lateral rather than vertical. Thus, for our first example, imagine a length of partly transparent tall fence with a hungry animal on one side and its food on the other. While a chicken would vainly keep on trying to fly over, a dog will run to the end of the fence, turn around, and then run back to the food.

This homely chicken-and-dog example reminds us of the correspondence between the complexity of behavior and the complexity of the nervous system. The bare statement of such a correspondence is the flattest of platitudes, but the conjugacy principle may nevertheless suggest strategies for learning something about the mechanism of that correspondence. For instance, one may wonder about the role of bypass in an animal's planning of its future behavior. Also, one may ask about the relevant neural circuitry: Which parts of the brain are responsible for arranging simple or complicated bypass sequences? How localized is this ability? At what evolutionary stages will such ability appear or jump up? Is there any connection, formal or otherwise, between bypass and human development theories such as that of Piaget?

As was mentioned before, communications technology was the first domain outside mathematics in which bypass was noticed. Accordingly, the second example deals with telephony. Here the problem, let us call it W, is to project voice over a long distance. One applies first the transformation S of acoustic vibrations into a pattern of fluctuating current, then comes the middle part T in which the fluctuating current is transmitted by cable and amplified if necessary; finally the last part S^{-1} retransforms the current back into acoustic vibrations, that is, voice. If wireless transmission is to be used then the simple bypass STS^{-1} is extended in depth to $SS_1T_1S_1^{-1}S^{-1}$. Here the outer part $S...S^{-1}$ is the same as before. The middle part T of STS^{-1}, which was the transmission by cable, is now replaced by the inner bypass $S_1T_1S_1^{-1}$. S_1 is the change of the fluctuating current into another pattern of electromagnetic vibrations, radio waves, followed by beaming from the transmitter antenna. T_1 is the electromagnetic transmission by radio waves, and S_1^{-1} is then the pickup at the receiving antenna followed by the transformation back into fluctuating current. The formal passage from STS^{-1} to $SS_1T_1S_1^{-1}S^{-1}$ will be called, for obvious reasons, bypass stacking.

The simple bypass $W = STS^{-1}$ and the extended or stacked bypass $W = SS_1T_1S_1^{-1}S^{-1}$ may be compared with reference to the previous remark, following (C_2), on bypass and levels. In $W = STS^{-1}$ there are two levels, voice or acoustic, and electric; so this bypass operates as follows: voice level to electric level—transmission by cable at the electric level—electric level to voice level. That is, the outer part $S...S^{-1}$ shifts the telecommunication problem from the acoustic level to the electric level. In $W = SS_1T_1S_1^{-1}S^{-1}$ there are three levels: voice, electric, electromagnetic. Thus the extended bypass operates as follows: voice level to electric level—electric level to electromagnetic level—transmission by radio waves at the electromagnetic level—electromagnetic level to electric level—electric level to voice level. Or: the outer part $S...S^{-1}$ of $W = STS^{-1}$ shifts the telecommunication problem from the acoustic level to the electric level, and the outer part $S_1...S_1^{-1}$ of $T = S_1T_1S_1^{-1}$ shifts it further from the electric level to the electromagnetic level. But the same effect is accomplished in a single shift from the acoustic level to the electromagnetic level by the outer part $(SS_1)...(SS_1)^{-1}$ of the extended or stacked bypass. This leads to the general formula for bypass extension by stacking; such extension arises by substituting a whole new bypass for the middle part of another bypass, and the formula is

(C_3) if $W = STS^{-1}$ and $T = S_1T_1S_1^{-1}$ so that $W = SS_1T_1S_1^{-1}S^{-1}$, then $W = (SS_1)T_1(SS_1)^{-1}$.

In words, this states that if W is conjugate to T under S, and T is conjugate to T_1 under S_1, then W is conjugate to T_1 under SS_1. In the digging and tunneling terminology of (C_2), the preceding formula describes the deepening of the shaft S by another shaft S_1 if, for instance, the wall extends below the ground level.

It was said before that bypasses started with mathematics, moved on to communications technology and then further to physics and other natural sciences, and finally were observed still further afield. The following third example was the first one to occur to the author as an instance of a possible bypass outside natural science or technology. In capitalist economics there arises the problem, call it again W_1, of operating without capital. Here the entrepreneur starts with the action S, which is borrowing the necessary capital, say from a bank. Then follows the middle part T_1; from the bank's point of view, T_1 is merely the term of the loan; from the borrower's point of view, T_1 means the production of goods and presumably profits by means of the borrowed capital. Finally comes S^{-1}, which is repaying the debt. This example may appear forced, but it illustrates very simply how bypasses combine in length. In the present example this refers to extending the loan for a further term T_2 under the formula

(C_4) if $W_1 = ST_1S^{-1}$ and $W_2 = ST_2S^{-1}$, then $W_1W_2 = ST_1T_2S^{-1}$.

Written out in full, the "product" W_1W_2 is $ST_1S^{-1}ST_2S^{-1}$ but $S^{-1}S$ cancels out since S and S^{-1} are inverses. This cancellation is here the formality of extending the loan for the further term T_2. With the digging and tunneling terms of (C_2), the formula (C_4) refers to combining two tunnels in order to get past two walls. The extension in depth also occurs in the present example. As before, such extension or stacking arises from a multiplicity of levels. Here these are the various possible credit levels, for instance, from a central state agency to a bank, and then to an individual borrower.

Mathematical examples of bypasses STS^{-1} will be given in which S and S^{-1} will be exact inverses and T will be smooth or loss-free. But unlike in mathematics, there is no perfect channel or transformation in nature or in society. Thus the last two examples of telephony and borrowing exhibit what might be called generalized friction: cable resistance and electrical noise in one case, interest on the loan in the other. Depending on circumstances and views, this is regarded as a nuisance, a necessary limiting factor, or a prime motivation. The whole aspect could be called the thermodynamics or entropy of bypasses in relation to losses, efficiency, and reversibility, and perhaps even to the price of novelty; this aspect will occupy us later.

Before going on to the next example the bypass rules (C_3) and (C_4) will be completed by adding the third and last rule, that for inverting a bypass:

(C_5) if $W = STS^{-1}$, then $T = S^{-1}WS$.

This states that if W is conjugate to T under S, then T is conjugate to W under S^{-1}. The following digging-and-tunneling illustration, somewhat fanciful but, hopefully, explanatory, can be given here. Imagine a hypothetical race of gnomes, dense and compact creatures who live underground, and move from A to B by tunneling. If in the process of motion a gnome meets a particularly refractory rock and wants to get past it—call his problem or difficulty T—he may make a bypass: make a shaft S^{-1} up from A to the surface, then execute a walk W on the surface of the ground to the point lying over B, and finally sink the reverse shaft S down to B. The gnome's bypass is schematized by $T = S^{-1}WS$.

The fourth example is military and concerns a novel type of tank used by one of the combatants in World War II. Since a ditch of width and depth comparable to the length of the tank will stop an ordinary tank, the following device was developed to perform the job W of getting across the ditch obstacle. The tank carried, somewhere, a pair of straight rails. On coming to the edge of the ditch, the action S was executed: lowering the rails into position and placing them across the ditch. The rest is obvious: T stands for driving across on the rails and S^{-1} for raising the rails and stowing them away in their resting position, prior to unobstructed rolling on.

This describes the discrete functioning of the modified tank in getting itself unaided across a ditch obstacle. However, the continuous functioning of an ordinary tank could also be expressed schematically in the form of an "infinitesimal" bypass as $\Delta W = \Delta S \Delta T (\Delta S)^{-1}$. Here ΔW is an element of forward motion, ΔS stands for unrolling an element of an iron "bridge" in front, ΔT is the element of forward translation on the just extended "bridge," and $(\Delta S)^{-1}$ means rolling up a "bridge" element behind.

The fifth example concerns only one object out of a large class that will be studied later. This class includes scaffoldings, splints, jacks, clamps, vises, and similar devices for temporary positioning, clamping, lifting, or holding. The bypass here consists of the usual three parts S, T, S^{-1}: S is putting the device on, T is holding it, and S^{-1} is taking it off. The common feature is that the important middle part T is passive, so that the things hold but nothing seems to be done, which is precisely the feature of usefulness. The same passivity will be observed later for other bypasses (e.g., those involving transfer or storage of energy or information, say in batteries or in computer memories). Also, something similar occurs in the commonest act under any economic system that has progressed beyond the barter stage: It is the bypass STS^{-1}, which starts with the selling S of one's goods or services for money; continues with T, which is keeping the money; and finishes with S^{-1}, which is buying the goods or services one requires.

Scaffoldings alone will be considered in the present example, partly to introduce the whole notion of "temporary holding devices" and partly to produce a nontechnical demonstration that the inverse of a product is the reversed product of inverses:

$$(S_1 S_2 \ldots S_n)^{-1} = S_n^{-1} \ldots S_2^{-1} S_1^{-1}.$$

This law of inverting noncommutative products is very well known to mathematicians and has an obvious bearing on multilevel bypass as in (C_3). Suppose then that a tower with n floors is being built, starting with the act of placing the first-floor scaffolding S_1 on the ground, then S_2 on S_1, S_3 on S_2, and so on. When S_n is in place, there follows the building T of the tower itself. Finally, the scaffolding units $S_1, S_2, \ldots, S_n$ are taken down in reverse order and the result is the bypass

$$(S_1 S_2 \ldots S_n) T (S_1 S_2 \ldots S_n)^{-1} = S_1 S_2 \ldots S_n T S_n^{-1} \ldots S_2^{-1} S_1^{-1}.$$

T itself could be broken down in the obvious way into a sequence of building successive floors:

$$T = T_1 T_2 \dots T_n.$$

Further, it may be assumed that the upper scaffoldings are not necessary to build the lower floors: only the part $S_1 S_2 \dots S_k$ is really needed to erect the kth floor T_k. There is now the bypass

$$S_1 S_2 \dots S_n T_1 T_2 \dots T_n S_n^{-1} \dots S_2^{-1} S_1^{-1}$$

with a certain restricted interchange, or commutativity, allowed:

$$S_i T_j = T_j S_i \quad \text{if} \quad i > j.$$

Certain questions could be raised now, some purely theoretical, others even of some practical interest with a view to optimal scheduling of unit operations. However, matters are rapidly growing technical, and so these questions will be deferred. But one remark is in order before moving on: Observe that the most universal of temporary holding devices is the hand. This observation takes us back to the first example. The questions asked there concerned the brain and bypass planning; now one could ask about the hand and bypass execution.

The sixth example might be called masking, blocking, or capping. It uses the stenciling idea and is the common schema for laboratory synthesis of certain chemical compounds (e.g., some polypeptides), for producing colored Easter eggs, and for batik dyeing. There are here the usual three parts of the bypass process: S, T, S^{-1}. The first one, S, does the protective masking. This may take the form of adding something to certain sensitive molecular groups so as to deactivate them temporarily. Or, in simpler cases, it may involve covering with wax a part of the egg or the material. T is the principal chemical synthesis or the sequence of such, or the dyeing process. Finally, the last leg of the bypass, S^{-1}, removes the protective wax coating or the protective added part of the molecular masking.

The seventh example suggests that certain features of humor seem to be related to bypass. For instance, an occasional form of a joke or a humorous story is one in which a bypass STS^{-1} is, or perhaps appears to be, executed, and then it is found that there was no wall to get past, or no middle part T, or no outer part S$\dots$S^{-1}. A specific instance of a connection between humor and bypass is the following theme of a folktale, a fable, or a comic poem, which occurs with some variants in several literatures. This concerns a hungry vagabond wandering through the countryside who manages to procure a free meal from a stingy though gullible peasant woman by promising to show her the cooking of a meal out of an iron nail. The performance starts with a pot of water over the fire and the nail is duly dropped in. The vagabond remarks that a handful of meal would make it faster and he gets it; then, to expedite the matters, a pinch of salt and a spoonful of butter are similarly procured. When the food is ready, the vagabond takes the nail out, eats, and the peasant woman is enchanted by the miracle.

The humor here might be granted to turn on the presence of something like bypass but it is still questionable what that something is. Naturally, one hesitates to speak of proofs or demonstrations in connection with humor; this would be impolite or, worse, clumsy. Even at the risk of such accusations another humorous story will be quoted which, if anything, is rather better known than the previous one. Here the element of humor arises from the undoubted presence of something like bypass stacked to several layers in depth. Of course, even the most punctilious rigorist in such matters might admit, without fearing any mathematical traps, that humor is a multilevel phenomenon. Or, inasmuch as humor concerns viewing a situation on several levels simultaneously, it might be said that humor is more of the nature of counterpoint than of melody.

The story concerns a poor farmer with a scolding wife, many noisy children, and a small hut. He complains about his troubles to his priest who has a great reputation for wisdom. On the condition of being strictly obeyed, the priest undertakes to ease the farmer's discomforts in six weeks, and at once orders the farmer to put the family goat into the hut. Next Sunday the farmer comes to the priest looking even more harassed, and he is told to take his dozen chickens into the hut as well. The week after he complains very bitterly but the priest only tells him to invite his mother-in-law to stay with him. This is the proverbial last straw: The next Sunday the farmer laments that life is not worth living and he is at the point of rebellion. Then the priest orders him to send the mother-in-law away; this eases the farmer's grievances considerably. The next week the chickens are put out, which continues the improvement; at the end of the sixth week, the goat gets expelled as well. Thereupon the farmer, smiling and content, thanks the priest and tells him that he has never realized how roomy and comfortable his hut is.

Since mathematics is universally known to be humorless, it may be disbelieved, even by mathematicians, that an element of the nail-in-the-pot fable could turn up in strictly professional mathematics, but here a reference may be made to the theorem of Thorin as expounded by Littlewood [89, pp. 21–22].

It is fitting to move from the ridiculous to the sublime, and so the next, eighth, example concerns that difficult branch of knowledge known as theology. This might begin with a brief reference to Hindu theology and, in particular, to the Hindu Trinity of Vishnu-Brahma-Shiva. From a functional point of view, already the names show a bypass somewhere in the background since the three deities are known as Vishnu the Creator, Brahma the Preserver, and Shiva the Destroyer.

Another Hindu concept, coming from the specific cult of Brahma, may also be mentioned. It is monotheistic in character and not consistent with the previous Trinitarian cosmology: the concept of Brahma Loka, or the Day of Brahma. This starts with the Cosmic Night and the Awakening of Worlds, expressed as the Awakening of Brahma, it continues with the Joyous Existence of Worlds, and it ends in the return to Cosmic Night or Brahma's Sleep.

It is fashionable nowadays to point out some enticing though possibly specious and facile parallels of Occidental science with Oriental religious thought; there is even a book [29] on the subject, titled *The Tao of Physics*. In this vein, it might be remarked that the aforementioned Hindu religious cosmology bears some resemblance to the scientific cosmology, in particular, to the "Big Bang" theory of the origin of the universe. The similarity would be even closer if the cosmological search for the missing matter (intergalactic gas? black holes? rest-mass of neutrinos?) were to be successful, for then the universe, after achieving maximum expansion, would start to contract, and would eventually finish, not with a whimper, but with a "Big Gnab." Thus the poet T. S. Eliot would be shown to be wrong here too. At any rate, the Brahma Loka is computed to last some four billion years, which is not too bad for a speculative approximation to the age of the universe.

The Judeo-Christian theology is more concerned with a sort of cosmic z-axis, a verticality, which is the moral dimension, rather than with expansion and contraction, or waxing and waning, as is the case in Hindu theology. This verticality shows itself even in the matter of terminology: Fall (or Sin) and Rise (or Redemption). To take one example, let us consider the Book of Job. The game that God and Devil play on Job is pure bypass in form: It starts with the successive inflictions on Job; it continues with Job's faith being proved to the uttermost; and it has a happy ending in the final blessings showered on Job (though nothing is said about Job's sons and daughters, who died so that a point could be proved). Yet, beside an element of a game or even a bet, this bypass includes a profounder element of what is known in theology as Amor Dei and even Amor Fati.

In all seriousness, and in all brevity, the Christian view of man's fall and post-fall condition might be schematized as STS^{-1}, where S is the Original Sin, T is the temporal existence in this world, and S^{-1} is Redemption. The symmetry of regarding Jesus Christ as the inverse to Adam and Mary as the inverse to Eve is no new thing in Christian theology. This symmetry extends even to such details as the attempt to show that Jesus expired in the ninth hour because that was the hour of the expulsion from Paradise. Also, there is the well-known observation that the most famous Marian hymn, "Ave Maria," starts with the word "Ave," which reverses to "Eva." An explicit statement of this symmetry of names occurs in the well-known anonymous ninth-century hymn "Ave Maris Stella" [117].

To speak of proofs and demonstrations would be yet harder here than in the previous example. However, even the matter of several levels and of bypass stacking by compounding in depth might be said to arise in theology. It will be recalled that the principal fountainhead of Christian theology is Saint Augustine, who came to Christianity from Manicheism by reading "certain books of the Platonists," as he records himself in the seventh book of his *Confessions*. Those books are believed to have been the works of Neoplatonists rather than of Plato, and perhaps even the *Enneads* of Plotinus himself. Now, in the Plotinian theology, the Fall and the Return occur through several levels of being called hypostases. The highest is the One, from there the descent is to the second one called the Mind, then to the third one called the Soul, and finally to the lowest one, which is the world of matter. The Plotinian redemption, described by Plotinus in a haunting phrase as the flight of the alone to the Alone, is just the reverse process of ascent back to the One.

Even if the theological bypass were granted, whether the Plotinian or the Christian one, it might still be asked what the obstacle or difficulty could be past which that bypass leads. This is a dark question, but a theologian's answer might hint at the problem of reconciling divine omnipotence with man's free will. This view could be certainly supported in the Augustinian case when we recall Augustine's battles with Pelagius; perhaps it could also be supported in the Plotinian case as well. As to the bypass of the Book of Job, the preceding view is perhaps the only one on which one could even attempt to justify the awful transgression done against Job.

A final remark related to the present example concerns a perverted use of bypass symmetry in the black arts. A black magician's work starts with a spell that invokes the demon. This is followed by the demon's performance of its task and then the spell of dismissal. It is interesting to observe a variant of the previously mentioned verticality here too: The demonic bypass first calls the demon up from some nether or infernal region; when the demon has done its work, it is dismissed back, or rather down, to its own region. The tragic consequences of the absence of the spell of dismissal are well known in traditional literature. Occasionally the bypass symmetry shows itself in the extremest form: The spell of dismissal is the repetition of the spell of invocation backwards.

Since the pole of the ridiculous and the pole of the sublime have both been touched upon, it may be that there remains for the ninth and last example only a sort of a halfway station between the two. Now, according to a well-known maxim of Napoleon, "from the sublime to the ridiculous there is only one step." But a half step from either pole toward the other one takes us to that most difficult of human endeavors, philosophy. Could bypasses be of any profitable concern at all to a philosopher?

Let us first glance very briefly at some episodes from the history of philosophy. There is Heraclitus with his emphasis on process as against permanence. One of his fragments runs: "All things exchange for fire and fire for all things, just as gold exchanges for goods and goods for gold." This generalization recalls something like bypass. Then there are certain forms of Greek terminology, such as for instance ἀναμνησις (recall or, literally, unforgetting) and ἀναγνωρισις (recognition, or literally, un-ignoring), used by Plato, Aristotle, and others. Their very structure suggests our inverse transformation S^{-1}. This is easily matched with a direct transformation S, and so there is the material for a bypass STS^{-1}. Passing over to Christian philosophy, we have Saint Augustine and his great concern with the problem of time and of measurement of time, from the eleventh book of his *Confessions*. It is interesting to speculate on the fact that no Pagan philosopher has raised similar question (with the possible exception of Plato's cosmic container—ὑποδοχειον), and on one of the principal formal, or descriptive, differences between Paganism and the monotheistic religions. The former is cyclic and repetitious, while the latter are linear and apocalyptic. Geometrically speaking, the model for the former is a circle; for the latter a line with a distinguished point (Covenant, Crucifixion and Resurrection, the Hegira). Now, it may be that the Augustinian problem of time and its measurement has a relation with the two bypasses: linear—periodic—linear and periodic—linear—periodic. The central problem of medieval philosophy, the nominalist-realist controversy on the status of individual and collective entities, has, in the author's opinion, an intimate though somewhat technical connection with bypass; this is discussed in Ex. 7.3. We cross finally to modern times and, having started with Heraclitus, we end up with Hegel. In view of the dynamic, almost dialectical, nature of bypass, this juxtaposition of the ancient dialectician with the modern one should not be surprising. It seems to the author that many aspects of bypasses have a distinctly Hegelian flavor. The principal dialectical triad of thesis, antithesis, and synthesis may be compared with the three basic bypass parts. There is in Hegel the ubiquitous idea of mediation, which seems to correspond to our bypass through a passage to another level. Hegel's mediation includes the important special case of self-mediation, which might be compared with our reflex bypass (to be mentioned soon). Even Hegel's "cunning of reason," described in his *Lectures on the Philosophy of World History*, a process whereby reason enlists passions in its interests, appears to have a distant connection with bypass.

The very considerable recent resurgence of interest in Hegel takes us right to the modern days. So we ask again: Aside from any possible historical explanatory interest, could bypasses be of any profitable concern to the philosopher? Of course, only the philosopher could answer that and it is probably easier for a mathematician to play the theologian than the philosopher. However, it is safe to assume that the philosopher would start his answer by asking some questions himself, perhaps such as

1. What do W, S, T, S^{-1} stand for?
2. In what sense are S and S^{-1} inverses?
3. How are S, T, and S^{-1} combined into STS^{-1}?
4. What does the equality sign in $W = STS^{-1}$ assert?

and other similar questions, all of them components in one large question: What is bypass? or to put it more personally to the author: What do you mean by bypass? This could be sidestepped by a countermove offered to the philosopher: You do not begin with definitions, why should I do so? In spite of an element of justice, this is a very poor gambit. It is probably improved by the remark that in an interdisciplinary domain (and new or old, bypasses are par excellence interdisciplinary), one is likely to do better by citing and examining many examples than by giving a few definitions. This explains the presence of the preliminary examples of this chapter, including the present one, and of the many examples to follow. In Chapter 3, an attempt will be made toward a better specification of what is meant by a bypass and its parts. But all this will at best only counter a few superificial objections without satisfying even the author. How could it satisfy the philosopher and tempt him to plow his way through the sequel? Obviously a far better lure must be presented.

Briefly, the best that can be done is this. The challenge to explain the meaning of bypass to the philosopher's satisfaction will be declined, and an offer will be made in its place. Instead of explaining bypass by semantics it will be attempted to explain semantics as bypass. If fulfilled in whole or in part, this offer may attract linguists and therefore philosophers as well. The linguistic-philosophical hook is baited openly, though clumsily. Many of the conjugacy examples, given already or to come, refer to transport or communication. That which is transported may be structure, as in the mathematical examples of bypass in Part 2, or it may be matter or energy or information, or even credit or power. We shall argue that a symbolizing communication process, such as language, is also mediated through a bypass, though a very complex one. It is tempting to say that if language communication is indeed through a bypass, then it is meaning that is transported, so that there is here something like the information theory or secrecy bypass of encode—transmit—decode. The crudest form of the language bypass would then be thoughts into words—transmission of words—words into thoughts. With two levels only, of thoughts and of words, this is very crude indeed but, as always with bypasses, there arises the possibility of trying to identify other intermediate levels. There is even the attractive idea of using some of those hypothetical intermediate levels to investigate that which the modern linguists of certain schools call deep structure of language, as distinguished from its superficial structure. Of course, some of the current problems of philosophy turn up at once because of the difference of the meaning received, the meaning transmitted, and the meaning intended.

But even if it is granted that the conjugacy principle has some bearing on language, something central is still missing and the possibility that bypass may supply this missing element is perhaps the strongest argument in our favor. Presumably, an adequate theory of meaning must be a court of last appeal: There is little use in trying to explain meaning if the explanation requires an explanation. This postulate of ultimateness is justifiable on psychological as well as on logical grounds: We wish to avoid the horror of infinity, or rather the horror of infinite regress, with the attendant dreaming or waking condition of mirrorless beholding of oneself, the back or the front of one's head included. It is clear what is wanted—some sort of power of high order that makes the system reflexive or self-referential or self-explaining. One is disposed to think that only very complex systems could have it. However, let us consider our fourth example. The tank in that example does not cross the ditch by any external means, bridge or whatnot; it manages to get itself across by itself. On rough terrain, it does not move over any external bridge; *it carries its bridge with itself*. So here is the crudest and most superficial example of reflexness achieved by the simplest application of bypass.

A better though more technical example of self-reference will be offered in Chapter 7. This concerns the use of generating functions to solve simple recurrence relations. These latter relate each member of an infinite sequence of numbers to some previous members (except for the first few, which are given). The bypass here starts with uniting the infinitude of members of the sequence at one level into a single object (the generating function) at another (collective) level. The recurrence relation, relating some members to *other* members, mirrors at the collective level into a functional relation relating the generating function to *itself*. Next, this generating function is found; finally comes the last step of the bypass in which the single collective object is separated back into an infinitude of individual ones, leading to the solution.

A much subtler reflex example is of course well known: the Gödel incompleteness theorem with its formula which asserts its own unprovability. A whole separate section will be devoted to a discussion of Gödel's theorem in the framework of the conjugacy principle. Somewhat as the bypass of the formulation (C_2) operates between the ground level and the underground level so, it will be suggested, Gödel's construction operates between the metamathematical level and the mathematical level. Here again reflexness comes from bypass.

In view of the preceding sketch, it is very tempting to try to explain, or rather to initiate an effort to explain, such profound matters as the reflex nature of meaning in language and the reflex nature of consciousness itself by an analysis involving the conjugacy principle. This is one of the deeper aims of the present book.

Chapter Two

Mathematical
Introduction

Conjugacy is very well known in mathematics but the unity of the phenomenon of conjugacy is obscured by the variety of names it bears in different branches of mathematics: similarity (in matrix algebra), integral–transform method (in functional equations), transformation principle (in geometry), and even conjugacy (in group theory and in iteration). It is characterized by the equation $W = STS^{-1}$, which defines W to be conjugate to T under S. An alternative way to state the conjugacy of W to T under S is to write $W \sim _S T$ or, if S does not matter, $W \sim T$.

For instance, let all capital letters stand for $n \times n$ matrices, and let the concatenation ST be the usual matrix multiplication. If S is nonsingular, let S^{-1} be its inverse. If $W = STS^{-1}$, then W and T are similar matrices. It follows that, for a positive integer $m, W^m = ST^m S^{-1}$. Thus, if n is at all large and T is diagonal or nearly diagonal, this affords a better way of computing the powers of W than the straightforward matrix multiplication.

With integral transforms W may stand for the problem of solving a (difficult) functional equation for an unknown function f. S corresponds to taking an integral transform of f, and T denotes the problem of solving the (easy) functional equation satisfied by the transform of f. Finally, S^{-1} means returning to the function domain from the transform domain: The inverse transform is applied to the solution of the (easy) transformed equation, yielding the unknown function f.

In geometry, W may be some unobvious proposition, or perhaps the problem of proving such a proposition, say about ellipses. S may be then a projection that reduces W to the special, and perhaps obvious, case T for circles. The bypass is completed by S^{-1}, whose meaning is clear. Or W may refer to a general conic, S may be a suitable projective transformation, and T may be the special case when the conic becomes a circle or degenerates to a pair of lines. Finally, W might be a problem in plane geometry, concerned with tangent circles and lines, and S might be the inversion in a suitable circle.

In iteration, one meets the problem of producing an explicit expression for the nth iterate W_n of a function W where, for simplicity, n is a positive integer. This problem arises often in connection with obtaining good nonlinear estimates for solving a nonlinear equation by successive approximations. Let all capital letters denote functions (of one variable), let multiplication ST mean functional composition, and let S^{-1} be the function inverse to S. Suppose that $W = STS^{-1}$, where T is some easily iterable function such as $x + 1$, cx or x^a. Then T_n is easily obtained and so the iteration problem is solved, since $W = STS^{-1}$ implies $W_n = ST_nS^{-1}$. These examples go some way toward justifying the formulations (C_0), (C_1), and (C_2) of the preceding chapter.

Conjugacy is an equivalence relation in the standard sense: It is reflexive, symmetric, and transitive. Now, the formula (C_3) in the preceding chapter is

$$\text{if}\quad W \sim_S T \quad\text{and}\quad T \sim_{S_1} T_1, \quad\text{then}\quad W \sim_{SS_1} T_1$$

so that (C_3) is merely a detailed statement of transitivity. Similarly, (C_5) is

$$\text{if}\quad W \sim_S T, \quad\text{then}\quad T \sim_{S^{-1}} W$$

so that (C_5) is just a detailed statement of symmetry. (C_4) states the cancellative multiplication rule

$$\text{if}\quad W_1 \sim_S T_1 \quad\text{and}\quad W_2 \sim_S T_2, \quad\text{then}\quad W_1 W_2 \sim_S T_1 T_2.$$

Thus the formal statements of the formulas (C_3), (C_4), and (C_5) are mathematically simple to the point of triviality. However, the three formulas themselves might be highly nontrivial from the point of view of applications of bypass, especially to domains which are neither mathematical nor easily mathematicizable. But there is also a considerable variety of applications of the conjugacy principle within mathematics itself, because of its many subdivisions if for no other reason, as will be shown in Chapters 4–7.

It is well known that mathematics may be divided up in many ways, for various reasons, and to serve diverse interests. There are the usual dichotomies: higher and lower, pure and applied, mainstream and offbeat, useful and useless, beautiful and useful. There are some simple and apparently natural subdivisions such as the chronological one into ancient, classical, and modern or the instructional one into elementary, introductory, intermediate, advanced, and research level. Of course, there is the standard division into fields such as algebra, analysis, combinatorics, geometry, logic and foundations, number theory, and so on, leading up to a full taxonomy by further splitting into subfields, sub-subfields, etc.

It was mentioned at the beginning of Chapter 1 that the work reported upon in this book had started as an informal attempt [96, especially Appendix] to isolate and describe some working principles of mathematics. Initially this was intended with a view to a yet different subdivision of mathematics: by working principles. It was hoped that by cutting across the boundaries of standard fields such a subdivision by principles would be not only different, but even differently different. How "differently"? The activity of dividing has two opposite or inverse activities: multiplying and unifying. Now, the intended subdivision by principles was meant to be unifying rather than multiplying or proliferating, as seems to be the case with usual subdivisions. However, the methodological interest in subdividing was soon lost, while the working principles themselves became more and more fascinating (and most of the fascination eventually settled on what we call the conjugacy or bypass principle).

Without requiring an exact rigorous answer, it might be asked: What is meant here by a working mathematical principle? Two immortal platitudes, one by Turan [135] and the other by Banach [136, p. 203] could be rephrased in partial answer: "a trick which has acquired dignity and currency by working often and widely" and "that which shows not only analogies between mathematical objects but also analogies between analogies." Because of its iterative element, the Banach quotation is reminiscent of the description of analogy by W. S. Jevons [77, p. 627]: "It has been said, indeed, that analogy denotes a resemblance not between things but between the relations of things." A serious and satisfactory answer to the last question is not easy. Presumably such an answer would have several components: One of them is width, as played up by Turan; another one is depth, as stressed by Banach.

On purely subjective and arbitrary grounds, the working principles, or rather the candidates for such, were classified with respect to some imagined weight as light, medium, and heavy. Some examples for each category will now be given; these examples appear here for several reasons. The first one is to provide a proper starting place for bypass among other principles. Next, some of the examples chosen will be helpful later in discussing connections between bypass and other things. Also, our examples provide some background against which to judge the value and scope of conjugacy, at least in its role as a mathematical principle. Finally, there is a certain intrinsic interest.

An example from the class coded privately as light is provided by the principle which asserts, loosely speaking, that multiplication is reducible to squaring: in symbols,

$$ab = \tfrac{1}{2}[(a + b)^2 - a^2 - b^2],$$

where a and b are complex numbers. Closely related is the fact that, under suitable conditions, the norm induces scalar product:

$$(A,B) = \tfrac{1}{2}[\|A + B\|^2 - \|A\|^2 - \|B\|^2].$$

At the trifling level of mathematical puzzles there is the following instance. Let $a_1, a_2, \ldots, a_k$ be k positive integers written in the ordinary decimal form, and suppose it is desired to find their squares $a_1^2, a_2^2, \ldots, a_k^2$ by multiplication. How *few* multiplications of integers are needed? The answer is 1, not k. For, a single "long" integer A is formed by stringing out the a's separated by suitable blocks of 0's:

$$A = a_1 0 \ldots 0 a_2 0 \ldots 0 a_k.$$

A^2 is obtained by a single squaring. Then A^2 yields not only the squares $a_1^2, a_2^2, \ldots, a_k^2$, spaced by groups of 0's, but also all the products $2a_i a_j$, $i \neq j$ as well. The puzzle becomes somewhat less trifling if each integer a_i has at most d digits, and it is asked how "short" A could be, without "spillover" occurring in A^2.

A less trifling application arises in number theory, with quadratic residues [137]. Let p be an odd prime, and for an integer a, define the Legendre symbol $\left(\dfrac{a}{p}\right)$ by

$$\left(\frac{a}{p}\right) = 1 \qquad \text{or} \qquad \left(\frac{a}{p}\right) = -1$$

according to whether a is, or is not, a quadratic residue mod p. That is, $\left(\dfrac{a}{p}\right)$ is 1 or -1, depending on whether the congruence

$$x^2 \equiv a \pmod{p}$$

does, or does not, have integer solutions x. In proving the complete multiplicativity law

$$\left(\frac{a}{p}\right)\left(\frac{b}{p}\right) = \left(\frac{ab}{p}\right),$$

it is simply shown that the product of two residues is a residue and the product of a residue and a nonresidue is a nonresidue: $R_i R_j = R_k$, $R_i N_j = N_k$. The remaining case is to show that a product of two nonresidues is a residue: $N_i N_j = R_k$. Although this is slightly harder to prove, it may be proved by reducing the multiplication $N_i N_j$ to squaring as follows. Let N_1 be a fixed nonresidue; if $R_1, R_2, \ldots, R_{(p-1)/2}$ are the distinct residues then the numbers $N_1 R_1, N_1 R_2, \ldots, N_1 R_{(p-1)/2}$ are incongruent nonresidues. Hence every non-residue may be written as $N_1 R_i$ and so

$$N_i N_j = N_1 R_i N_1 R_j = N_1^2 R_i R_j = R_k,$$

since N_1^2, like any perfect square, is a quadratic residue by definition.

A serious application of the multiplication-is-reducible-to-squaring principle arises in Mikusinski's proof [99] of the Titchmarsh convolution theorem: If

$$f*g = \int_0^x f(y)g(x-y)dy$$

and all functions are continuous, then $f*g \equiv 0$ if and only if $f \equiv 0$ or $g \equiv 0$. This theorem is fundamental in Mikusinski's theory of distributions. The theorem asserts that the ring of continuous functions, under ordinary addition and convolution-multiplication, has no zero divisors; therefore, it can be embedded in a field of quotients, and these quotients include distributions. The original proof of Titchmarsh was complicated, and it was simplified by Mikusinski as follows. First, he proves the "square case":

$$f*f \equiv 0 \longrightarrow f \equiv 0.$$

Once this is shown, using an ingenious elementary reduction, the general or "multiplication case" of the Titchmarsh theorem, that is,

$$f*g \equiv 0 \longrightarrow f \equiv 0 \qquad \text{or} \quad g \equiv 0,$$

is deduced from the special "square case" mentioned earlier.

Although multiplication ab is reducible to squaring, the product abc is not expressible by cubes. There is a much weaker formula here:

$$ab(a+b) = \tfrac{1}{3}[(a+b)^3 - a^3 - b^3],$$

which is occasionally useful. For instance, let I be a real open interval of length c and let $x_1, x_2, \ldots$ be any sequence of distinct real numbers in I that is dense in I. Removing x_1 from I breaks it up into two open intervals I_1 and I_2. Then x_2 is removed, which breaks up one of them into two new open intervals, and so on. Eventually, every interval I_n that appears so in the process gets broken up into two parts; let their lengths be u_n and v_n. Now a simple exercise in cancellation, together with the preceding formula, shows that

$$\sum_{n=1}^{\infty} u_n v_n (u_n + v_n) = \frac{c^3}{3}.$$

The next example in the light class is a combinatorial principle that might be called parity counting, residual counting, or counting by correspondence. This concerns finite sets, and the idea is to count by avoiding counting, or to reduce difficult counts to easy ones, by employing a 1:1 correspondence as far as possible. The object may be to show that two finite sets A and B are of equal size (without needing that size), or to show that they must be unequal, or to find the difference $A - B$ of two large and nearly equal numbers, or to count A by establishing a 1:1 correspondence with B and counting B instead of A.

A very simple application shows that the symmetric group has as many even permutations as odd ones, since a 1:1 correspondence between them is established by a fixed transposition. Another simple use is in the solution of a well-known puzzle: Let two squares lying on any 45° diagonal line be removed from an ordinary chessboard, and suppose that a domino piece is a rectangle that just fits over two neighboring chess squares. Can the mutilated chessboard be covered exactly by 31 dominoes? No, because each domino must cover one black square and one white, whereas the two removed squares are necessarily of the same color.

An effective use of this principle is made by Euler to prove that the number of k-combinations out of n distinct objects, say out of $(1, 2, \ldots, n)$, is $\binom{n+k-1}{k}$ when repetitions are allowed. We recall that an ordinary k-combination, that is, one without repetitions, is any sequence

$$i_1, i_2, \ldots, i_k \qquad \text{such that} \quad 1 \leqslant i_1 < i_2 < \ldots < i_k \leqslant n$$

so that $0 \leqslant k \leqslant n$, and there are $\binom{n}{k}$ of them. A k-combination with repetitions is any sequence

$$i_1, i_2, \ldots, i_k \qquad \text{such that} \quad 1 \leqslant i_1 \leqslant i_2 \leqslant \ldots \leqslant i_k \leqslant n.$$

Euler shows that there is a $1:1$ correspondence between the k-combinations out of $(1,2,\dots,n)$ with repetitions, and the k-combinations out of $(1,2,\dots,n+k-1)$ without repetitions. This proves his formula, and the correspondence is simply established:

$$i_1,i_2,i_3,\dots,i_k \leftrightarrow i_1,i_2+1,i_3+2,\dots,i_k+(k-1).$$

A correspondence-counting argument may be used to prove the following theorem of Erdos–Szekeres: A sequence of $1+mn$ distinct real numbers contains either a decreasing subsequence of length $>m$ or an increasing subsequence of length $>n$. Let $\mathrm{Fal}(i)$ and $\mathrm{Ris}(i)$ be the lengths of the longest decreasing and increasing subsequences starting with the ith one of the terms $s_1,s_2,\dots,s_{1+mn}$ of the original sequence. Consider the correspondence

$$F:i \to \langle \mathrm{Fal}(i),\mathrm{Ris}(i) \rangle, \qquad i = 1,2,\dots,1+mn$$

between integers i and ordered pairs of integers. It is shown first that this is a $1:1$ correspondence:

$$i \neq j \text{ implies } \langle \mathrm{Fal}(i),\mathrm{Ris}(i) \rangle \neq \langle \mathrm{Fal}(j),\mathrm{Ris}(j) \rangle,$$

for $i \neq j$ implies that $s_i \neq s_j$, and it may be assumed that $i < j$. Next, if $s_i < s_j$, then $\mathrm{Ris}(i) \geqslant 1 + \mathrm{Ris}(j)$ since an increasing sequence starting with s_j gets extended to a longer increasing sequence by being prefaced with s_i. Dually, $\mathrm{Fal}(i) \geqslant 1 + \mathrm{Fal}(j)$ if $s_i > s_j$. Now the theorem is proved by contradiction: If the result were false then $1 \leqslant \mathrm{Fal}(i) \leqslant m$ and $1 \leqslant \mathrm{Ris}(i) \leqslant n$, but then F would have been a $1:1$ correspondence between a set of $1+mn$ elements and a set of $\leqslant mn$ elements. A related though stronger result is the theorem of Dilworth [39] on partially ordered sets P: The smallest number of disjoint chains covering P equals the largest number of pairwise incomparable elements.

Another result that can be simply proved with counting by correspondence is Cayley's theorem [103] on labeled trees. We recall that a labeled tree on n vertices is a connected cycle-free graph with n vertices labeled $1,2,\dots,n$; Cayley's theorem asserts that the number of such trees is n^{n-2} $(n \geqslant 2)$. The very form, n^{n-2}, raises the possibility of a proof by establishing a $1:1$ correspondence between the distinct labeled trees on n vertices and all ordered $(n-2)$-tuples $\langle a_1,a_2,\dots,a_{n-2} \rangle$ of integers a_i such that $1 \leqslant a_i \leqslant n$. Such a correspondence has been found by Prüfer [103]: Consider any such tree T; there must be in it a vertex of valency 1. Take the lowest-numbered such vertex and remove it from T together with its single edge. This erased edge leads to a vertex numbered a_1, which defines the first member of the $(n-2)$-tuple. The same operation is then repeated on the remaining tree till

$\langle a_1, a_2, \ldots, a_{n-2} \rangle$ has been obtained. The correspondence is shown to be $1:1$ by checking that the known $(n-2)$-tuple allows a complete restitution of the original tree T coded by it; this includes of course the labeling of T's vertices.

The same counting principle finds a frequent use in deriving various results for partitions. One of the best-known such arguments is the counting proof of F. Franklin [63, p. 284] of Euler's identity

$$\prod_{n=1}^{\infty} (1 - x^n) = 1 + \sum_{n=1}^{\infty} (-1)^n [x^{n(3n-1)/2} + x^{n(3n+1)/2}].$$

To begin with, it is observed that

$$\prod_{n=1}^{\infty} (1 - x^n) = 1 + \sum_{n=1}^{\infty} [pe(n) - po(n)] x^n,$$

where the partition functions $pe(n)$ and $po(n)$ are as follows: $pe(n)$ is the number of ways of representing n as a sum of an even number of distinct positive integers; $po(n)$ is the same but with an odd number of integers. In each case, the order of the summands is disregarded. The idea of the proof is to avoid counting the representations of n which make up the numbers $pe(n)$ and $po(n)$, as far as possible. Since only the difference $pe(n) - po(n)$ matters, a possibly complete cancellative $1:1$ correspondence will be established between the even and the odd representations, and only the cases not covered by that correspondence will be counted. This is done by using the so-called Ferrers graph, a geometrical aid that shows the representation

$$n = n_1 + n_2 + \ldots + n_k$$

graphically. Since the summands n_i are unequal and their order is irrelevant, it may be assumed that $n_1 > n_2 > \ldots > n_k$. Hence the preceding representation of n is given by an aligned array of n dots with n_1 dots in the first row, n_2 in the second row, and so on. A particular representation $14 = 6 + 5 + 3$ is shown in Figure 1a. The dots in the lowest row are called the base, those on the $45°$-slope line from the rightmost dot are called the slope. The same names are used for the corresponding numbers, so that in Figure 1a base $= 3$ and slope $= 2$. If the base and the slope are disjoint, then the slope may be moved over to form a new base (as from Figure 1a to Figure 1b), or the reverse operation may be attempted. The same is true even if they are not disjoint provided that the slope exceeds the base. The important thing is that both operations exchange the parity of the representation. The only exceptions, that is, the only cases when such parity exchange is impossible

(a)   (b)

Figure 1

in this scheme, occur when the base and the slope have a dot in common and either base = slope or base = 1 + slope. The proof of Euler's identity is easily completed by examining and counting those special cases.

The final illustration is the generalized Archimedes' theorem. This goes outside the discrete realm and concerns the continuous domain, but the proof is in exactly the same spirit as the previous ones. Instead of counting two finite sets A and B by means of a 1:1 correspondence, the interest is now in measuring certain two sets A and B, one on the (ordinary) sphere S and the other on the cylinder C which envelopes S. It is shown that A and B have equal areas because they can be decomposed into elementary parts of the same area. Let X be the common axis of S and C, and let p be a point on S other than the north and south poles in which X cuts S. The unique point in C, which is closest to p, is called the axial projection $\phi(p)$ of p. Let A be any sufficiently regular set on S and let B be the image on C of A under ϕ. Take on the cylinder C the obvious area element: a small curved rectangle R_1 bounded by two generators and two circles. Let R be the small curved rectangle on S whose image under ϕ is R_1. The name "rectangle" for R_1 and R is justified since all their angles are right. It is now verified that R and R_1 have the same area (in the limit) because on passing from R_1 to R one pair of "sides" is stretched in the same ratio in which the other pair is compressed. Going through the usual steps of setting up an integral, it is found that A and B have the same area. It is easily checked that there is no trouble with the exceptional points (the two poles). As a special case, A may be the whole of S and we get then the result of Archimedes: The area of a sphere of radius r is $4\pi r^2$. It may be observed that this Archimedean formula was to some extent anticipated by more than a millennium. Moreover, it is interesting to note just how it was anticipated: by a slight extension of the 1:1 correspondence. In a mathematical Egyptian papyrus, now in Moscow, it is stated that if a hemispherical basket and its circular plane lid are woven from reeds (and are of the same thickness), then twice as much reed material is needed for the basket as for the cover.

Three things will be mentioned in concluding this topic. First, our principle enters into the counting arguments based on the pigeonhole principle and, to a smaller extent, into the arguments of Ramsey's theorem [118]. Second, our principle is used in showing the equivalence of Lindelöf's hypothesis to a lattice-point counting problem [96]. Finally, this principle could be forced into a bypass form in which there are two domains A and B, and the counting starts with a map from A to B, continues by proceeding or arguing in B, and ends with a return from B to A. A repeated use of this form of bypass can be detected even for counting of infinite sets, for instance, in Bernstein's proof [16] of the Cantor–Schröder–Bernstein theorem: If A and B are two infinite sets each of which is in a $1:1$ correspondence with a subset of the other, then A and B are in a $1:1$ correspondence themselves.

The last example from the light class to be considered here is the principle of structural parameter; to honor Leonhard Euler who used it so signally, it could also be called Euler's principle. The idea is that in a given mathematical situation there may not be enough structure to begin with. So, a new parameter t is introduced, suitable operations involving t are carried out, and then t is "canceled" or "withdrawn" by being set equal to 1 or, occasionally, to some other special value. Applications are many since this is a real workhorse of elementary and other mathematics.

So, to evaluate a sum like $\Sigma n^k a_n$, the power series

$$\sum a_n t^n = f(t)$$

is introduced, the differential operation $t\,d/dt$ is applied k times, and then we set $t = 1$. To compute the infinite product,

$$\prod_{n=1}^{\infty} (1 - 2^{-n})$$

to some considerable accuracy (see Reference 95, p. 220), one starts with

$$\prod_{n=1}^{\infty} (1 - t^n),$$

then the already proved Euler identity is applied, and finally one puts $t = 1/2$, getting

$$\prod_{n=1}^{\infty} (1 - 2^{-n}) = 1 - 2^{-1} - 2^{-2} + 2^{-5} + 2^{-7} - 2^{-12} - 2^{-15} + 2^{-22}$$
$$+ 2^{-26} - 2^{-35} - 2^{-40} + \dots .$$

To prove Euler's theorem about the function $f(x,y,z,\dots)$ which is homogeneous of degree a, the definition of homogeneity is observed:

$$f(tx,ty,tz,\ldots) = t^a f(x,y,z,\ldots),$$

the operation d/dt is applied to both sides, and then t is set equal to 1, yielding the desired theorem:

$$x\frac{\partial f}{\partial x} + y\frac{\partial f}{\partial y} + z\frac{\partial f}{\partial z} + \ldots = af.$$

Taylor's expansion theorem for a function of several variables is

$$f(x+h, y+k, \ldots, z+l) = \exp\left(h\frac{\partial}{\partial x} + k\frac{\partial}{\partial y} + \ldots + l\frac{\partial}{\partial z}\right) f(x,y,\ldots,z);$$

this may be proved by considering the auxiliary function

$$f(x+ht, y+kt, \ldots, z+lt)$$

as a function of t alone, applying to it McLaurin's theorem, and setting $t = 1$. The Euler expansions

$$\prod_{n=1}^{\infty}(1 + x^{2n+1}) = 1 + \sum_{n=1}^{\infty}\left[x^{n^2}/\prod_{j=1}^{n}(1 - x^{2j})\right],$$

$$\prod_{n=1}^{\infty}(1 + x^{2n}) = 1 + \sum_{n=1}^{\infty}\left[x^{n(n+1)}/\prod_{j=1}^{n}(1 - x^{2j})\right]$$

can be proved similarly. We start by introducing Euler's parameter t and forming

$$F(t) = \prod_{n=1}^{\infty}(1 + tx^{2n+1}).$$

The replicating property of the factors leads to the functional equation

$$F(t) = (1 + tx)F(tx^2),$$

which in turn leads to the recursion

$$c_j = x^{2j-1}(1 - x^{2j})^{-1}c_{j-1}, \qquad j = 1,2,\ldots$$

for the coefficients in the power series for $F(t)$. This determines $F(t)$ as a power series, and the two Euler expansions are obtained as $F(1)$ and $F(x)$.

A discrete device closely related to Euler's principle is used to help obtain recursions in certain purely combinatorial computations. For instance, let $A_0 = A_1 = A_2 = 1$, and for $n > 2$ let Z_n be the number of zigzag permutations, introduced by D. André, [3]. Here a permutation $(k_1, k_2, \ldots, k_n)$ of $(1, 2, \ldots, n)$ is called a zigzag if no three consecutive neighbors k_i, k_{i+1}, k_{i+2} are either in rising or in falling order of size. By taking complements $(k_i \to n + 1 - k_i)$ it is observed that there are as many initially rising zigzags $(k_1 < k_2)$ as initially falling ones $(k_1 > k_2)$ and as many finally rising zigzags $(k_{n-1} < k_n)$ as finally falling ones $(k_{n-1} > k_n)$. In particular, Z_n is even: $Z_n = 2A_n$. It does not appear easy to get a recursion for Z_n directly. So, consider a zigzag in which the largest number n occupies the $(r + 1)$st place, $0 \leqslant r \leqslant n - 1$. By adding this structural parameter r, the desired recursion is simply obtained because any finally falling zigzag on the preceding r elements combines with any initially rising zigzag on the $n - r - 1$ following elements, to form a full n-zigzag. Conversely, any full n-zigzag decomposes so, for some r. This leads to the recursion

$$2A_n = \sum_{r=0}^{n-1} \binom{n-1}{r} A_r A_{n-1-r}.$$

If

$$y = \sum_{n=0}^{\infty} A_n x^n / n!$$

is the exponential generating function, then the previous recursion shows that

$$1 + y^2 = 2y'$$

so that $y(x) = \tan(x/2 + \pi/4)$. Therefore,

$$\tan x = A_1 x + A_3 x^3 / 3! + A_5 x^5 / 5! + \ldots,$$

$$\sec x = A_0 + A_2 x^2 / 2! + A_4 x^4 / 4! + \ldots.$$

It must be observed finally that the Euler principle of structural parameter seems to fit well into the bypass framework STS^{-1}: Introduce the structural parameter t, perform the necessary t-operations, remove t. There is even a certain functional similarity, perhaps farfetched but not entirely imaginary, to the fifth example from Chapter 1 concerning "temporary holding devices." In particular, the structural parameter t plays a temporary but essential role similar to that of a scaffolding used in the construction of a tall structure.

Only one example from the middleweight class will be given here but that single example—the principle of computing or expressing one thing in two different ways (and equating the results)—is a host in itself. To begin with, there is a preliminary observation that will be taken up later in this section, and then again in discussing bypass and novelty: The present principle deals with identities; yet, or rather therefore, it has some fundamental bearing on mathematical novelty. This remark might be developed a little further. Misled by a misunderstanding of the sign of equality perhaps, some post-Wittgensteinian philosophical reactionaries may dissolve mathematics, to their own satisfaction at least, into a vast system of tautologies, $a = a$. But even a most superficial use of the two-ways principle produces novelties. Thus representing a number in two different ways as

$$x^2 - y^2 = (x - y)(x + y)$$

shows something immediate, informative, and otherwise unobvious: An integer that is a difference of squares of two integers differing by more than 1 cannot be prime.

To continue with some very simple instances of the two-ways principle, there is the following proof of the integral formula

$$\int_0^\infty \frac{\sin x}{x} dx = \frac{\pi}{2}.$$

This begins by considering the double integral

$$\iint e^{-xy} \sin x \, dA$$

over the positive quadrant; the preceding integral formula results at once by evaluating and equating

$$\int_0^\infty \int_0^\infty e^{-xy} \sin x \, dy dx \quad \text{and} \quad \int_0^\infty \int_0^\infty e^{-xy} \sin x \, dx dy.$$

Other simple applications may be obtained from a large collection of combinatorial counting arguments. Suppose, for instance, that a committee of n persons is to be chosen out of a total of $2n$ people: n women and n men. A direct "sexless" count gives $\binom{2n}{n}$ as the number of different choices. A committee of k women and $n - k$ men can be chosen in

$$\binom{n}{k}\binom{n}{n-k} \quad \text{or} \quad \binom{n}{k}^2$$

ways; summing over k and equating the sum to the direct count gives the identity

$$\sum_{k=0}^{n}\binom{n}{k}^2 = \binom{2n}{n}.$$

It is interesting to note that the usual proof of this identity seems different, but it also uses the two-ways principle: the coefficient of x^n is computed in two ways, from the two sides of the identity

$$(1+x)^n(1+x)^n = (1+x)^{2n}.$$

Another combinatorial counting instance is in obtaining the formula

$$\sum_{j=1}^{n} j^2 = \frac{1}{6}n(n+1)(2n+1).$$

To prove this, consider a regular tetrahedral array consisting of n layers of identical balls in contact; there is one ball in the top layer, three in the next one, six in the next, and so on. There are

$$\sum_{k=1}^{j} k = \frac{j(j+1)}{2}$$

balls in the jth layer from the top, and so a total of

$$S = \frac{1}{2}\sum_{j=1}^{n} j^2 + \frac{1}{2}\sum_{j=1}^{n} j = \frac{1}{4}n(n+1) + \frac{1}{2}\sum_{j=1}^{n} j^2$$

balls in all. The preceding count is, geometrically, vertex-to-base. Let us do the count again, but edge-to-opposite-edge. This time the jth layer is a plane rectangular array of balls, of size j by $(n+1-j)$. Therefore, the total is

$$S = \sum_{j=1}^{n} j(n+1-j) = (n+1)\sum_{j=1}^{n} j - \sum_{j=1}^{n} j^2 = \frac{1}{2}n(n+1)^2 - \sum_{j=1}^{n} j^2.$$

Equating the two counts, we get the desired formula for the sum of squares.

A combinatorial-algebraic instance is the class-equation of a finite group G [68, 142]. Here the elements of G are counted once directly and once by conjugacy classes. Equating the counts gives the class-equation

$$\text{ord } G = \sum_a \text{ord } G/\text{ord } N(a)$$

where $N(a)$ is the normalizer of a in G and the summation runs over any set of elements with exactly one element in each conjugacy class.

It may happen that the unique answer to a physical problem is obtainable in two mathematically different ways; we can then equate the answers. For instance, let an electrical point-charge e be placed inside a plane domain D which is either a rectangle or its limiting case: an infinite or semi-infinite rectangular strip. The interest is in finding the potential distribution inside D when the boundary of D is kept at potential 0. This problem can be solved by means of a suitable conformal mapping which maps D onto the upper half-plane; the mapping function is elliptic (for the rectangle) or its limiting case (exponential or trigonometric, for the strips). Another way of solving is by the method of images: A complete system of images of e is produced in the lines bounding D, signed $+$ and $-$ in alternation, and the potential due to such a singly or doubly infinite system of images gives the answer. Equating the two solutions, one gets a series expansion for the mapping function.

In the continuous domain, computing in two ways appears in the Cavalieri principle, a precalculus method for calculating volumes of solids named after B. Cavalieri (1598–1647): If two (ordinary three-dimensional) solids can be placed so that every plane perpendicular to a fixed line cuts them in sections of equal area, then the two solids have equal volumes. For a use of an extension to n dimensions, see [98]. In the modern framework of measure theory, the Cavalieri principle relates to the well-known theorem of Fubini [62] on product measures. This asserts that the integral of an integrable function of two variables can be evaluated to the same value by integrating in either of the two orders.

An important special subcase of the two-ways principle arises when one function is expressed in two different ways. A further sub-subcase might be called the $\Sigma\Pi$-principle: when one expression is a series (i.e., sum) and the other a product. Many special applications occur by using the uniqueness theorem for the series in question and equating the coefficients in the two-way identity. Here again identities produce novelty, even if it is just the novelty of the way of expressing a simple mathematical fact. Usually this fact is the uniqueness of a certain representation. Several instances will be given.

Let $A = a_1, a_2, \ldots$ be a sequence, finite or infinite, of increasing positive integers with $a_1 = 1$, and let k be a positive integer $\geqslant 1$. The partition function

$p(n,A,k)$ is defined as the number of representations of $n(n \geqslant 1)$ as a sum of certain a_i, order disregarded, so that no summand a_i occurs more than k times. Here the $\Sigma\Pi$-identity

$$\prod_{n=1}^{\infty} \frac{1 - x^{(k+1)a_n}}{1 - x^{a_n}} = \sum_{n=1}^{\infty} p(n,A,k)x^n$$

defines the ordinary generating for $p(n,A,k)$. Important special cases arise when A is the sequence of natural integers $1,2,3, \ldots$ or a subsequence of it that forms an arithmetic progression, and $k = 1$ or $k = \infty$.

The function $(1 - x)^{-1}$ may be expressed as the usual geometric series and also by the Euler product

$$\frac{1}{1 - x} = \prod_{n=0}^{\infty} (1 + x^{2^n}).$$

Equating the coefficients in this $\Sigma\Pi$-identity expresses the unique representation of a positive integer as a sum of powers of 2. A more general $\Sigma\Pi$-identity

$$\sum_{n=0}^{\infty} x^n = \prod_{n=0}^{\infty} \frac{1 - x^{b^{n+1}}}{1 - x^{b^n}}$$

leads in the same way to the uniqueness of the representation of any n in the scale of b.

The Riemann function $\zeta(s)$ may be expressed as a Dirichlet series and also as the Euler product over the primes; the $\Sigma\Pi$-identity is

$$\sum_{n=1}^{\infty} n^{-s} = \prod_{p=2}^{\infty} (1 - p^{-s})^{-1}.$$

Comparing the coefficients leads to the fundamental theorem of arithmetic: Every positive integer > 1 is a unique product of prime powers. The preceding theorem extends to Riemann functions of algebraic fields other than the field of rational numbers; the product on the right-hand side has then the factors $(1 - [N(p)]^{-s})^{-1}$, where $N(p)$ is the norm of the prime ideal p, and the product extends over prime ideals.

Finally, let p be a prime and let $I(n)$ be the number of irreducible monic polynomials of degree n over the Galois field F with p elements. Then the $\Sigma\Pi$-identity

$$\frac{1}{1 - px} = \prod_{n=1}^{\infty} (1 - x^n)^{-I(n)}$$

also expresses the uniqueness of a certain representation: A monic polynomial over F is a unique product of irreducible monic polynomials over F. This is because the coefficient of x^d in the series for $(1 - x^n)^{-I(n)}$ is the number of those monic polynomials of degree d that are products of the $I(n)$ irreducible ones of degree n, and no others. This follows from the binomial series and from Euler's expression for combinations with repetition. Thus the right-hand side of the $\Sigma\Pi$-identity enumerates the products of powers of irreducible monic polynomials. That is, it enumerates the polynomials by degree; hence it equals the left-hand side, which enumerates the same directly. Logarithmic differentation and the Möbius inversion formula lead to

$$I(n) = \frac{1}{n} \sum_{d|n} \mu\left(\frac{n}{d}\right) p^d,$$

from which it follows by an easy estimation ($\mu(1) = 1, \mu(k) \geqslant -1$) that $I(n) > 0$ for every $n \geqslant 1$. That is, for every $n \geqslant 1$, there exist irreducible monic polynomials of degree n. Since the residue classes modulo such a polynomial form a Galois field with p^n elements, it follows that such fields exist for every $n \geqslant 1$.

A rather different use may be made of the two-ways principle by expressing in two ways a function $f(z)$ of a complex variable. It may happen that one expression, though perhaps less familiar, is valid in a domain wider than that of the other well-known expression. Then one side of the two-ways identity provides an analytic continuation of the other side. Several such examples exist, for instance, the Borel and the Mittag–Leffler summability methods [38].

The two-ways principle is concerned in the theory of equivalence of sets by finite decomposition. Let X and Y be two sets in some Euclidean space E^d, and let there be defined a certain collection of sets in E^d, which are called elementary parts. Suppose that

$$X = \bigcup_{i=1}^{n} U_i, \qquad Y = \bigcup_{i=1}^{n} V_i$$

are disjoint decompositions of X and of Y, each into n elementary parts. Suppose also that for $i = 1, \ldots, n$, U_i is congruent to V_i by a rigid motion. Then X and Y are said to be equivalent by finite decomposition: $X \approx Y$. The relation $X \approx Y$ may be expressed as an instance of the two-ways principle: There are n elementary parts which can be reshuffled in two ways, once to form X, then again to form Y. Further, $X \approx Y$ could be written schematically in the bypass form

$$X = (STS^{-1})Y.$$

Here S stands for the operation of decomposing into a disjoint finite union of elementary parts, T stands for the operation of reshuffling, and S^{-1} for the operation inverse to S: taking a union of disjoint elementary parts. Two examples of such equivalence by finite decomposition are particularly well known. The first one is Hilbert's theory of area of plane polygons [46, Ch.5; 69], which uses no limits. Here the sets X and Y are plane polygons and the elementary parts are triangles. Since areas only matter, two triangles are called disjoint if they share no interior points. The principal theorem asserts that $X \approx Y$ if and only if area X = area Y. The other example is the Hausdorff–Banach–Tarski paradoxical decomposition of the sphere, [11]. Here X, Y, and the elementary parts are arbitrary subsets of the ordinary two-dimensional sphere S, and the congruence of elementary parts is by a rotation of S. Using the axiom of choice, it is shown that $S = X \cup Y$ where the disjoint sets X and Y are such that $X \approx S$ and $Y \approx S$.

As with several other principles, either discussed already or yet to come, bypass seems to be hovering somewhere in the background. One rather obvious possible connection with the two-ways principle hinges on regarding the conjugacy-defining equation $W = STS^{-1}$ itself as a two-ways identity. Both the matter of bypass levels and the matter of novelty (if indeed they are two rather than one) appear to be concerned. In a very simple formulation, we start with an initial position (or configuration, or condition, or situation) P_0 and a final one P_1. Let P_0 and P_1 be in a certain suitable sense at the same level A; the interest is in a passage from P_0 to P_1. An obvious way of doing it, call it W, is to proceed from P_0 to P_1 at the level A. But on account of obstacles, walls, or barriers this may be unobvious, difficult, or (we should like to add) perhaps impossible. The novelty of doing it in another way starts with realizing that there may be another level, say B. Now the passage is by the usual bypass recipe: From P_0 in A to Q_0 in B along S, then from Q_0 to Q_1 in B along some path T in B, and finally from Q_1 in B to P_1 in A along S^{-1}. Thus the bypass equation $W = STS^{-1}$ appears as a two-ways identity. Another connection of bypass with the two-ways principle starts with two bypasses $W = STS^{-1}$ and $W = S_1 T_1 S_1^{-1}$ and the two-ways identity is $STS^{-1} = S_1 T_1 S_1^{-1}$. An instance of this, concerning the use of different generating functions, is discussed at some length in Chapter 7.

Finally we come to the last class, which contains the heavyweight principles. Three examples will be taken up: reductio ad absurdum, induction, and the principle of stratification into layers by increasing complexity.

The first one is also known as the method of indirect proof or of proof by contradiction. It depends on the law of excluded middle, the *tertium non datur*, according to which a proposition P is true if it is not true that

P is not true. No instances need be given here—the principle is absolutely fundamental and universally used in usual mathematics at all levels. Its universality becomes quite apparent when we compare the richness of usual mathematics with the poverty, virtuous intellectually though it might be, of the mathematics of intuitionists and other constructivists who reject this principle. It is interesting to note that a proof by this method runs like this: It is false that the proposition is false. If D stands for the denial and P for the proposition then the preceding phrase might be put into the bypass from DPD^{-1}. To establish this, it would be necessary to show that in some sense the denial D is its own inverse: $DD = 1$. This is the case if the identity 1 is taken as the affirmation: $D^2 = 1$ is then the law of excluded middle. Now, this binary disjunctive (it is tempting to say: Boolean) nature of D is precisely what the intuitionists deny. But it must be noted that they deny it only for infinite domains. The last remark will turn out to be important: It will show that in one form or another all our three heavyweight examples concern infinity. Here "infinity" might perhaps be replaced by "infinity and closure."

To sum up, the method of reductio ad absurdum may be regarded by a classical nonintuitionist mathematician as a round trip between the levels of truth and of falsehood. He starts at the level T and a hypothetical assumption takes him to the level F; a contradiction is derived then, which purchases the return ticket and takes him back to T, completing the bypass and the proof.

Before leaving the subject of the excluded middle, we must remark that this principle is not really restricted to logic and mathematics. On the contrary, it might be argued that an odd form of it, very far removed from logic or mathematics, appears often in personal or intimate conversations between old acquaintances, friends, or lovers. We refer here to such modes of address as "dear sir," "madam," "you clod," "you half-wit," "you hag," "you old bag," "you pig," "you dog," "you dirty old man," "you son of a...," "you ... old broad," as used between people who admire, respect, or even love each other. It is interesting to note that the closer the relation, the more seemingly insulting words may be used. This suggests that the whole thing may be an odd sort of a tie-emphasizing device to combat loneliness and separateness. There may also be present in it an element of wishing to avoid hubris: It is bad and even dangerous to heap praise or compliments on those one likes or loves. However, it would take us too far afield to discuss those possibilities here. Purely formally, the whole phenomenon may be schematized as follows: The intention of the one is to call the other nice, clever, wonderful, near, beautiful or some such thing; the first operation is to deny it, and so the opposite is expressed, with the complete understanding of its being denied again by the other. Thus the bypass of the excluded middle serves here to help transport sentiments of intimacy past the wall of separateness.

On the less, but not too much less, personal basis something similar seems to operate in the literary form of composition known as satire. The author wishes to affirm that something is bad (or, less often, good), and again the first action is to deny it. Thus the opposite idea is expressed and it is taken for granted that the second denial, which completes the bypass, is the reader's. The need for an active audience participation flatters and otherwise emphasizes the effect.

We come now to induction. Its best-known form is what might be called linear induction; the name is justified by the linear order with which it deals and on which it depends. There are two forms of linear induction: short and long. In the short case the domain is the positive integers, 1,2,3, ... (or perhaps the nonnegative integers 0,1,2, ...). The following simple picture illustrates this form of induction. Let a row of identical bricks stand on a horizontal plane, each brick standing on its smallest face, let the row start with a distinguished initial brick, and let the separation between each two neighbors be constant and small. If the initial brick is pushed over onto its neighbor, then the whole row falls.

In the long case, the domain 0,1,2, ... is stretched further to include transfinite ordinals, and the process is then called transfinite induction. It could even be claimed with some justice that transfinite ordinals might be profitably regarded as infinite integers, and that they are constructed so as to make long induction work. Both the short and the long form exhibit a strong element of directionality: from smaller numbers to larger ones. This directionality comes from the linear order, and it is rather more pronounced in the long, or transfinite, case: Every ordinal, finite or infinite, has a unique successor (and that is, roughly, why linear induction works) but not every ordinal has a predecessor. Such "non-isotropy of the before and the after" is perhaps one of the main stumbling blocks to a student beginning the subject. Some of those matters will be discussed further in the section on sets, ordinals, cardinals, and patterns in Chapter 7.

A simple form of short linear induction concerns a proposition $P(n)$ about a positive integer n. It is supposed that

(1) (a) $P(1)$ is true, (b) $P(n + 1)$ is true whenever $P(n)$ is true.

The inductive conclusion is that $P(n)$ is true for all positive integers. Here positive integers could be replaced by nonnegative ones, and (a) is then changed to $P(0)$ is true. Part (a) of (1) may be called the initialization, and part (b) the inductive glue. In this form induction is used to prove many propositions, for instance, that the sum of the cubes of first n positive integers equals the square of the sum of those integers.

There is also a different form of induction in which (1) is replaced by

(2) (a) $P(1)$ is true, (b) $P(n + 1)$ is true whenever $P(1),P(2), \ldots, P(n)$ are all true.

and it is inductively concluded that $P(n)$ is true for all positive integers. This form is sometimes more convenient to use than (1), for instance, in an inductive proof of the unique prime-power factorization of positive integers. It is useful to have names to distinguish (1) and (2): (1) may be called successor induction, (2) may be called continuation induction. The continuation form has the merit of extending at once to the long, or transfinite, induction: let $P(\alpha)$ be a proposition about an ordinal α and suppose that

(3) (a) $P(0)$ is true, (b) $P(\alpha)$ is true whenever $P(\beta)$ is true for all $\beta < \alpha$;

the transfinite-inductive conclusion is that $P(\alpha)$ is true for all ordinals α.

We note next an outline of the construction of ordinals in an axiomatic set theory, such as the common form of Zermelo–Frankel, as modified by von Neumann. Zero is defined as the empty set ϕ with no elements, 1 is defined as the set $\{\phi\}$ with the single element ϕ, 2 is the set $\{\phi,\{\phi\}\}$, and so on, so that each ordinal after 0 is defined as the set of all its predecessors. The phrase "and so on" must be examined further. The nonlimit ordinals (i.e., ordinals with a predecessor) such as 7, $\omega + 1$, or $\omega^2 + 5$ are defined in one way. Limit ordinals (i.e., ordinals without a predecessor), such as the first infinite ordinal ω, are defined in another way. To put it roughly, the nonlimit ordinals are defined as in the successor form of induction: by throwing in one more. The limit ordinals are defined as in the continuation form of induction: by collecting together all that precedes them. This technique of definition of ordinals is justified in our set theory: Both "adding 1" and "collecting together all that precedes" are legitimate processes allowed by the axioms. This brief outline may be taken as a partial justification of a previous claim that the ordinals are constructed so as to make long induction work.

The preceding outline also emphasizes something very well known: Induction may be used to define or to construct as well as to prove. With this purpose, it is often called recursion, especially in its short form. Thus a function $f(n)$ may be given for all nonnegative integers n by a simple successor recursion:

$$f(0) \text{ is given}, \quad f(n + 1) = G(f(n)) \quad \text{for} \quad n = 0,1,2, \ldots$$

where G is known. This generalizes to a multi-successor recursion: k is a fixed positive integer, the initializing values $f(0),f(1), \ldots, f(k)$ are given, and the inductive (or recursive) glue is

$$f(n + k + 1) = G(f(n + k), f(n + k - 1), \ldots, f(n)), \qquad n = 0,1,2, \ldots$$

where G is a known function of $k + 1$ variables. For instance, the Fibonacci numbers $F_0, F_1, F_2, \ldots$ are given recursively as

$$F_0 = F_1 = 1, \quad F_{n+2} = F_{n+1} + F_n, \qquad n = 0,1,2, \ldots .$$

All that has been said so far concerns linear induction; we take up next another type which might be called multiple or n-dimensional or, preferably, patterned induction. It shares with short induction the property of operating on an infinite countable domain D, but this infinitude D has a structure different from linear order. The concern here is mostly with propositions about several integers or with functions of several integer-valued variables. So, the domains D of interest are integer-lattice domains: Infinite subsets of the lattice L_d of all points in the d-dimensional Euclidean space, every one of whose d coordinates is an integer (negative, 0, or positive). However, more general domains could be considered: subsets of other lattices or (the vertex sets of) certain infinite directed graphs. Suppose then that D is a given infinite subset of L_d, usually, though not always, the nonnegative octant of L_d or a set congruent to it. Next comes the initialization: A nonempty subset I of D is prescribed. Finally, the inductive glue must be specified. In the linear case, this was done in terms of the successor structure of ordinals, and so linear induction proceeds by means of nextness. Here the inductive glue is expressed in terms of a related but more general notion, and patterned induction proceeds by a spreading contagion. In more detail, the contagion is a rule for extension or adjunction, under which new points in D correspond to certain finite collections of points in D. The problem of patterned induction is to test the strength of the inductive glue: to decide whether the smallest subset of D, containing I and closed with respect to the given extension rule, is D itself.

A simple example, which is probably closest to the linear induction, is as follows. The finite dimension d is arbitrary, D is the nonnegative integer lattice octant

$$D = \{(x_1, \ldots, x_d) : x_i \text{ is an integer} \geq 0 \text{ for } i = 1, \ldots, d\},$$

and I is the origin $(0, \ldots, 0)$ alone. Given two points p and q in D, a majorization relation $p \prec q$ is defined, quite analogous to linear order: Let

$$p = (p_1, \ldots, p_d), \qquad q = (q_1, \ldots, q_d),$$

then $p \prec q$ if and only if

$$p_i \leqslant q_i \qquad i = 1, 2, \ldots d,$$

and at least one of the d inequalities is strict. Now the extension rule is given by making every point q in D correspond to the finite collection

$$C(q) = \{x: x \in D, x \prec q\}.$$

Given a subset X of D, the set

$$N(X) = \{x: x \notin X, C(x) \subset X\}$$

may be called the set of neighbors of X. It is now a simple matter to verify that if we start with I, adjoin its neighbors, then the neighbors of the enlarged set, and so on, we get all of D. That is, the only subset of D, containing I, which cannot be extended by the preceding adjunction rule is D itself. The situation has some analogy to starting with a given functional element in the complex plane and continuing it to the domain of holomorphy of the function by means of power-series continuation. Another observation is that patterned induction is concerned with closure, and this reminds us that even linear induction, in its transfinite form, can be disguised in terms of maximal or closure principles (such as Zorn's or Tukey's).

We return now to our simple example. Many further problems suggest themselves here; we mention one only. Let the contagion rule for adjunction be fixed, for instance, as it was earlier. We ask: What is the geometrical characterization of all subsets D_1 of D that are finitely initializable? That is, we wish to characterize geometrically all subsets D_1 in D that possess a finite initializing subset I_1 and are such that the smallest subset of D_1 that contains I_1 and is closed under this contagion rule is D_1 itself.

A new variant of patterned induction arises when the contagion rule is probabilistic rather than deterministic. This leads to the percolation processes introduced by Hammersley and co-workers (see Reference 51 and the references given there). For a specific two-dimensional example, let D be an infinite subset of the plane integer-lattice, let I be a single point in D, and let a simple inclusive contagion rule be by spreading to nearest lattice-neighbors. Suppose that D is connected: Any two points in it can be joined by a chain of nearest neighbors in D. Then it is clear that the smallest subset of D, containing I and closed under this adjunction rule, is D itself. Now reformulate the problem probabilistically: All conditions are as before except that the contagion has a fixed given probability $p(0 < p < 1)$ of spreading to any immediate neighbor of an "infected" point, independent of what happens elsewhere. Suppose for simplicity that D is the whole lattice. What is the "right" question to ask? Obviously not about the spread of infection to the whole of D. But recall that the purpose of any induction is, so to speak, to produce an infinity by finite means. Thus the "right" question to ask is about the critical value of p so that the infection spreads infinitely (i.e., does not remain contained in any finite subset of D). As Hammersley shows, such a critical value exists for D.

As an example in patterned induction, let us consider Ramsey's theorem [118]. There are two forms of it, for infinite sets and for finite sets; we begin with the infinite case, which is the simpler one. Let S be any infinite set; without loss of generality S may be taken as the set of positive integers 1,2, ... Let r and m be any two positive integers. Consider the set S_r of all r-combinations of S, that is, of all unordered r-tuples of elements in S (or briefly, of r-tuples in S). Let S_r be split in any way into m disjoint sets $A_1, \ldots, A_m$. Then Ramsey's theorem asserts that S contains an infinite subset X all of whose r-tuples lie in a single set A_i. The finite form is obtained on replacing infinite sets by finite ones; the conclusion is then that if S is large enough, it contains an arbitrarily large subset X all of whose r-tuples lie in a single set A_i. To state it in detail, a third parameter n is needed: the size of the set X. The theorem is then as follows. Let r, m, n be any positive integers, with $r \leqslant n$. Then there exists a positive integer $f = f(r,m,n,)$ with the following property: let $S = \{1,2, \ldots, f\}$, and let the set S_r of r-tuples of S be split in any way into m disjoint sets $A_1, \ldots, A_m$; then there is a subset X of S, with n elements, such that all r-tuples of X lie in a single set A_i.

Given a classification of r-tuples of $S = \{1,2, \ldots, f\}$, a subset X of S is called homogeneous if all its r-tuples fall into a single class of r-tuples. Now the sense of the preceding theorem may be briefly explained thus: If an a priori fixed classification is given for the r-tuples of a growing set S, then there will appear arbitrarily large homogeneous sets. This connects the finite and the infinite cases of the theorem and shows how the former implies the latter.

The strategy of the proof will be sketched next. We want to show that the (smallest possible) integer $f = f(r,m,n)$ exists for all r, m, n. There are some obvious reductions: It may be assumed that $m > 1$ and $n > r$ since trivially $f(r,1,n) = n$ and $f(r,m,r) = s$, where s is the smallest integer for which $\binom{s}{r} \geqslant m$. Thus the first nontrivial case of Ramsey's theorem concerns the value of $f(2,2,3)$, which turns out to be 6. This is known as the problem of six people: Among any six individuals, there are either some three, every two of whom know each other, or some three no two of whom know each other. The value 6 cannot be reduced as is shown by the example of five people sitting around a small circular table, among whom every two neighbors know each other but no others do. A generalization of the problem of six people concerns the existence of $f(2,2,n)$. This may be simply phrased in the language of graph theory: For every positive integer n, there is a unique smallest positive integer

$f(n) = f(2,2,n)$ such if G is any graph on $f(n)$ vertices, then either G or its complement $\bar{G}$ contains a complete n-graph.

It may be shown by ordinary (i.e., short linear) induction that the preceding proposition about graphs implies the full Ramsey theorem. Hence there is the possibility of proving Ramsey's theorem by applying an ordinary induction to show the existence of $f(n)$ for all n. However, the situation here is as before with the principle of structural parameter: After the reduction to a single variable n, there is not enough structure in the problem to give an (easy) inductive hold. Thus it is preferred to treat the three-variable case and prove the existence of $f(r,m,n)$ for all positive integers r, m, n such that $m > 1$ and $n > r \geqslant 1$. First, a moderately simple combinatorial argument [109, p. 43] shows that

$$f(r,2m,n) \leqslant f(r,m,2^{f(r,m,n)}), \qquad m > 1. \tag{4}$$

Since $f(r,m,n)$ is clearly nondecreasing when m grows, the case of general $f(r,m,n)$ follows by ordinary induction on m from the case of $f(r,2,n)$. Therefore, it remains to show that $g(r,n) = f(r,2,n)$ is defined for all integers r,n such that $n > r \geqslant 1$. In terms of Ramsey's theorem, the number $g = g(r,n,)$ is as follows: If $S = \{1,2,\ldots,g\}$ and the collection S_r of all its r-tuples is split into two disjoint classes, then S contains a subset with n elements, all of whose r-tuples lie in just one of the two classes. Again, there is not quite enough structure in it for an easy inductive hold, and so an extra parameter is introduced. As will be seen, this extra parameter converts a symmetric situation with less structure into a dissymmetric situation with more structure. Instead of $g(r,n)$, one considers a slightly more general quantity $h = h(r,p,q)$ for which the following will be proved: If $S = \{1,2,\ldots,h\}$ and the collection S_r of r-tuples in S is split into two disjoint classes A and B, then either some subset of S with p elements has all its r-tuples in A or some subset of S with q elements has all its r-tuples in B. Clearly

$$g(r,\min(p,q)) \leqslant h(r,p,q) \leqslant g(r,\max(p,q)),$$

so that the existence of the dissymmetric form with $h(r,p,q)$ implies the existence of the symmetric form with $g(r,n)$ (and conversely).

Everything is finally ready for an application of a patterned induction: to show that $h(r,p,q)$ is defined for all positive integers r, p, q such that $q \geqslant r \geqslant 1$, $p \geqslant r \geqslant 1$. The dimension d is 3 and the domain D is the following subset of the integer lattice in E^3:

$$D = \{(r,p,q): p \geqslant r \geqslant 1, q \geqslant r \geqslant 1\}.$$

To initialize our patterned induction, we observe that $h(r,r,q) = q$ and $h(r,p,r) = p$ trivially. Also, for $r = 1$, the pigeonhole principle shows that $h(1,p,q) = p + q - 1$. These special cases give us the initializing subset I of D

$$I = \{(r,r,q): q \geqslant r\} \cup \{(r,p,r): p \geqslant r\} \cup \{(1,p,q)\}.$$

Finally, another moderately simple combinatorial argument [109, p. 46] shows that

$$h(r,p,q) \leqslant 1 + h(r - 1, h(r,p - 1,q), h(r,p,q - 1)), \tag{5}$$

which provides the necessary inductive glue. Now it can be checked that $h(r,p,q)$ is defined for all points (r,p,q) in D: For $r = 1$ from I, then for $r = 2$ [by initializing at $(2,2,q)$ and $(2,p,2)$ and applying (5)], then for $r = 3$ in a similar way and so on.

A detailed examination shows that the use of patterned induction in the preceding example, as in many other similar ones, can be replaced by a number of applications of ordinary [i.e., linear and (usually) short] induction. Thus it may be asked: Is patterned induction really necessary? That is, can every instance of it be resolved into a number of applications of ordinary induction, or are there inductive proofs or inductive constructions for which it is unavoidable? A second question that might occur to the reader is: Why is so much space devoted to, or wasted on, Ramsey's theorem? This is because of a mathematical discovery made in 1977 [13, p. 1133]: Using an apparently slight modification of Ramsey's theorem, mathematicians have succeeded for the first time in producing a reasonably natural combinatorial proposition about integers, which is true but unprovable in elementary arithmetic. The mere existence of such true but improvable propositions has been known since Gödel's work (1931) on incompleteness. This is discussed in a separate section of Chapter 7, where the role of bypass is considered in producing self-referentiality. In the meantime, we formulate this version. First, the Ramsey theorem itself will be restated:

for all positive integers r,m,n $(r \leqslant n)$ there exists a positive integer $f = f(r,m,n)$ such that if $S = \{1,2, \ldots, f\}$ and the set S_r of all r-tuples in S is split into any m disjoint classes, then S contains a homogeneous subset X of size at least n.

It will be noticed that the only change from the previous form is the inessential substitution of "size at least n" for "size n." To formulate the new version, which is due to Paris and Harrington [13], we need first a simple definition. Consider any initial segment of the set of positive integers, that is, a set $S = \{1,2, \ldots, \phi\}$, where ϕ is a positive integer. Then a subset X of S is called relatively large if and only if the number of elements of X is at least as big as the smallest member of X: card $X \geqslant \min X$. Now the true but unprovable version of Ramsey's theorem is

Theorem (Paris–Harrington): for all positive integers r,m,n ($r \leqslant n$) there exists a positive integer $\phi = \phi(r,m,n)$ such that if $S = \{1,2, \ldots, \phi\}$ and the set S_r of all r-tuples in S is split into any m disjoint classes, then S contains a homogeneous subset X of size at least n, which is relatively large.

The sense of this theorem can be briefly explained in the same way in which Ramsey's theorem itself was explained before: If an a priori fixed classification is given for the r-tuples of a growing set S, then it is impossible to prevent the occurrence of a homogeneous subset of S that is both large and relatively large.

We return now to the linear and the patterned mode of induction and recursion. The problem, as always with the induction principle, is how to prove an infinitude of statements, or define an infinitude of values, or construct an infinitude of objects, or compute an infinitude of numbers, by finite means. There is no essential difference between proving or constructing, as with induction, and defining or computing, as with recursion. A useful neutral word is "producing"; this includes proving, defining, constructing, and computing. So, briefly put, the problem of induction is how to produce an infinitude by finite means. The very idea of the induction principle is to start the production going, and to make it continue, or propagate, by itself (very much like Disraeli's "machines making machines" in the quotation that follows). For linear induction there are no difficulties, either with the start or with the continuation. Both are given "naturally": the one as the distinguished first element (0,1, or the empty set), the other from the linear order (i.e., "nextness"). For patterned induction, both the start and the continuation are prescribed: the one as the initializing set, the other in the form of a continuation rule.

Perhaps the best general illustration of what is behind induction that can be given at all, is given in English prose, and not in mathematical terminologizing. It is a sentence from *Coningsby*, a novel by Benjamin Disraeli written in 1844 [40, Book 4, Ch.2], and it describes the impact of industrial machinery on a young upper-class Englishman who is entirely innocent of any previous knowledge of industry or machines: "And yet the mystery of mysteries is to view machines making machines; a spectacle that fills the mind with curious, and even awful, speculation." This antedates by 15 years Darwin's *On the Origin of Species* (1859), by 28 years Samuel Butler's "Book of the Machines" in his *Erewhon* (1872), by 34 years George Eliot's essay "Shadows of the Coming Race" in her *Opinions of Theophrastus Such* (1878), by 47 years the appearance of Giuseppe Peano's axioms for natural numbers and the explicit statement of (short linear) induction (1891), and by more than a century the self-replicating automata of John von Neumann (1948–1966). Any one of these names could be chosen as a direction in which to interpret Disraeli's dark sentence. Whichever name and direction is taken, the idea of induction, that is, of an infinitude achieved by finite means, suggests itself.

Working with recursion rather than induction, we shall now attempt to show how and why the linear mode is insufficient and the patterned mode is indispensable. The main object of recursion theory is to formalize the intuitive notion of computing effectively an infinity of values or, in more detail, of computing effectively functions $f(x_1, \ldots, x_n)$. Here $x_1, \ldots, x_n$ are nonnegative integer variables and the values of f are also nonnegative integers. One introduces first the class PR of primitive recursive functions. This class contains all the initializing functions of the form

$$f(x_1, \ldots, x_n) = c \qquad \text{(constant functions),}$$
$$f(x_1, \ldots, x_n) = x_i, \ 1 \leqslant i \leqslant n \qquad \text{(projections),}$$
$$f(x) = x + 1 \qquad \text{(Peano successor function).}$$

Also, the class PR is (minimally) closed under two schemas: substitution and induction. The substitution schema is:

if $f(x_1, \ldots, x_n)$ and $g_1(x_1, \ldots, x_k), \ldots, g_n(x_1, \ldots, x_k)$ are in PR then the function

$$f(x_1, \ldots, x_k) = f(g_1, \ldots, g_n)$$

is also in PR. The induction schema, which is also called the schema of primitive recursion, is:

if $f(0, x_2, \ldots, x_n)$ and $g(x_0, x_1, \ldots, x_n)$ are in PR, and if

$$f(x_1 + 1, x_2, \ldots, x_n) = g(f(x_1, x_2, \ldots, x_n), x_1, x_2, \ldots, x_n)$$

then $f(x_1, \ldots, x_n)$ is in PR. It can be shown that most of the functions ordinarily used are in fact primitive recursive: $x + y$, xy, x^y, $n!$, the remainder after dividing x by y, the nth digit in the decimal representation of π, the nth consecutive prime, and so on.

The induction schema certainly embodies in a generous form the successor type of induction expressed in (1): reducing $f(n + 1)$ to $f(n)$. But it might be suspected that the continuing form of induction, expressed by (2), could perhaps lead outside the class PR: reducing $f(n + 1)$ to the whole course of previous values $f(n), f(n - 1), \ldots, f(0)$. Of course, a definition of the type

$$f(n + 1) = H(f(n), f(n - 1), \ldots, f(0))$$

is not admissible since functions with a growing number of variables are not allowed. However, there are legitimate ways in which $f(n + 1)$ could still depend on its whole previous history. Suppose, for instance, that $f(0)$ is given and

$$f(n + 1) = h(\max(f(0), \ldots, f(n))),$$

where $h(x)$ is in PR. Is f in PR? We show first that

$$U(n) = \max(f(0), \ldots, f(n))$$

can be handled by primitive recursion. It is verified that the function $\max(x,y)$ is in PR, and the induction schema is used to define

$$U(0) = f(0), \qquad U(k) = \max(f(k), U(k - 1)).$$

Hence $f(n)$ is in PR. What about other more complicated cases of f? For that matter, what does it mean to say that $f(n + 1)$ depends "legitimately" on all its previous values? To answer these questions, we use the standard encoding

$$S(n) = \prod_{i=0}^{n} p(i)^{f(i)}$$

where $p(0) = 2$, $p(1) = 3$, ... and $p(i)$ is the ith prime. Since $p(i)$ is in PR, and so is the maximum function, we can recover from $S(n)$ the values of $f(n)$ and also of $f(n - 1), \ldots, f(0)$, by the unique prime-power factorization. Thus any primitive recursion by the continuing form, on $f(n), f(n - 1), \ldots, f(0)$, reduces to primitive recursion on $S(n)$ with n. Briefly put, nothing new is gained by continuing induction that could not be obtained already by successor induction.

It turns out that the class PR is too narrow to represent all effectively computable functions. One method to show this is to enumerate by an effective procedure all one-variable functions in PR as a sequence $f_p(n)$, $p = 0,1,\ldots$. Then the Cantor diagonalization is used to define the function $\delta(n) = f_n(n) + 1$. Since the enumeration is effective, $\delta(n)$ is effectively computable and obviously not a member of PR. A method that is perhaps more illuminating, is to produce an effectively computable function $A(n)$ which grows faster than any primitive recursive function $f(x_1, \ldots, x_k)$:

$$A(x_1 + x_2 + \ldots + x_n) > f(x_1, \ldots, x_n), \quad x_1 + \ldots + x_n \text{ large enough.}$$

Such a function $A(n)$ is produced by patterned induction; since linear induction does not take us outside PR, the existence of $A(n)$ substantiates the claim that linear induction is insufficient and patterned induction is therefore necessary.

The class PR of primitive recursive functions is first enlarged to the class R of recursive functions. This can be done in a variety of ways, such as Turing's rigorization of machine computability, Markov's theory of algorithms, calculi of the so-called λ-conversion, and general recursion theory; in each case the class R of functions declared to be effectively computable is the same. A convenient way to extend PR to R is due to Kleene [80]: The class R contains the same initializing functions as PR, it is (minimally) closed under the same schemas of substitution and primitive recursion as PR, and also under the additional schema of minimalization: if $f(x_1, \ldots, x_n, x)$ is a function in R such that for any set of values $x_1, \ldots, x_n$ a value of x exists for which $f(x_1, \ldots, x_n, x) = 0$, then the function

$$g(x_1, \ldots, x_n) = \min\{x : f(x_1, \ldots, x_n, x) = 0\}$$

is itself in R. An examination of the way in which R is set up shows that it is eminently reasonable to regard the functions in R as effectively computable; the converse idea, that every effectively computable function is recursive, i.e., in R, is known as the Church Thesis [80].

The notion behind producing a recursive function $A(n)$ which grows faster than any primitive recursive function is simple and depends, informally speaking, on suitably exploiting iteration. The idea is due to D. Hilbert [70], the details and construction are given in W. Ackermann [2], and a certain simplification has been achieved by, and is given in exhaustive detail in, H. Hermes [67]. In terms of a simple computing schema for PR, functions such as $x + k$ are produced without loops (i.e., without using induction schema), functions such as $x + y$ are produced by a one-loop program (i.e., with one use of induction), functions such as xy are produced by a program with a loop within a loop (i.e., with two uses of induction), and so on. Generally, each function in the sequence

$$x + y, \; xy, \; y^x, \; y^{y^{\cdot^{\cdot^{y}}}} \quad (x + 1 \text{ occurrences of } y), \ldots$$

is in PR but it calls for one further depth of loop-embedding, or one more application of induction, than the previous function. Also, it is observed that each function, considered as a function of y alone, increases faster than any bounded iterate of the preceding function. Therefore, to produce $A(n)$, one starts by defining the preceding sequence $f_n(x,y)$ of functions in PR so that $f_{n+1}(x,y)$, as a function of y, grows faster than any bounded iterate of $f_n(x,y)$, and then one diagonalizes: $A(n) = f_n(n,n)$. The definition of $f_n(x,y)$ is by a single defining relation and suitable initializations.

A certain simple but essential bypass may be noted here:

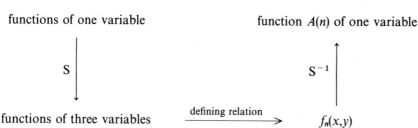

functions of one variable

function $A(n)$ of one variable

S

S^{-1}

functions of three variables $\xrightarrow{\text{defining relation}}$ $f_n(x,y)$

We start with one-variable functions in PR, and we wish to majorize them all at once. First, the domain is extended to three-variable functions by the extension S, then the defining relation T is used to define $f_n(x,y)$ (this T turns out to be a double recursion not reducible to a single recursion or, in our terms, a patterned induction not replaceable by a linear induction), and finally we "contract" from the three-variable function $A(n)$ by the process of diagonalization, which in this context appears as S^{-1}.

A certain analogy may be observed here to what might be called the Apollonian bypass (Ex.4.9). Namely, Apollonius is interested in investigating the properties of conics, which are plane curves, by the synthetic method of Greek geometry. To do this, he first takes his plane and embeds himself in the three-dimensional space, then he intersects there various surfaces (cones, planes, spheres), and finally he reenters the plane, having proved some specific property of a conic. However, the need for the Apollonian bypass (plane to space—work in space—space to plane) is eliminated by Descartes. On the other hand, the preceding bypass for obtaining $A(n)$ is not eliminable.

The definition of $f_n(x,y)$ is by the initializations

$$f_0(x,y) = x + y, f_{n+1}(0,y) = \begin{array}{ll} 0 & \text{if } n = 0 \\ 1 & \text{if } n = 1 \\ y & \text{if } n > 1, \end{array}$$

and the single defining relation

$$f_{n+1}(x + 1,y) = f_n(f_{n+1}(x,y),y). \tag{6}$$

Therefore, $f_n(x,y)$ is defined for all nonnegative integer values n,x,y by patterned induction, and it is shown that $f_n(x,y)$ is recursive [67, 80]; hence $A(n)$ is also recursive. For the first few values of n, we have

$$f_0(x,y) = x + y, f_1(x,y) = xy, f_2(x,y) = y^x,$$

$$f_3(x,y) = y^{y^{\cdot^{\cdot^{\cdot^y}}}} \quad (1 + x \text{ occurrences of } y), \quad \text{and}$$

$$f_4(x,y) =$$

$$y^{y^{\cdot^{\cdot^{\cdot^y}}}}_{x} (\# \text{ of } y\text{'s is } 1 + y^{y^{\cdot^{\cdot^{\cdot^y}}}} \;(_{x-1} \# \text{ of } y\text{'s is } 1 + y^{y^{\cdot^{\cdot^{\cdot^y}}}} \;(_{x-2} \# \dots 1 + y^{y^{\cdot^{\cdot^{\cdot^y}}}} \dots$$

$$(\# \text{ of } y\text{'s is } 1 + y) \;) \dots)$$
$$_1 \qquad\qquad\qquad\qquad _1 \; _2 \quad _x$$

where the subscripts enumerate the left and right brackets and pair them together.

The preceding development may be connected with the previously discussed theorem of Ramsey. The crucial size-determining estimate there was the inequality (5) for $h(r,p,q)$. Replace r,p,q by $r + 1, p + 1, q + 1$, write $h_r(p,q)$ for $h(r,p,q)$ and replace the inequality sign by an equality, for bounding purposes. Then (5) becomes

$$h_{r+1}(p + 1,q + 1) = 1 + h_r(h_{r+1}(p,q + 1),h_{r+1}(p + 1,q))$$

which is very similar to (6). It might even be hypothesized that the Ramsey function $f(r,m,n)$, like the Ackermann function $A(n)$, grows too fast to be primitive recursive. Finally, let us recall the Paris–Harrington function $\phi(r,m,n)$ from the Paris–Harrington unprovable version of Ramsey's theorem. This function ϕ cannot be recursive for otherwise, by standard techniques, the theorem could be shown to be not only true but also provable in Peano arithmetic. Now, the Ackermann function $A(n)$ was constructed by the double recursion (6), which cannot be reduced to a single recursion. By exercising ingenuity it is possible likewise to construct, for any positive integer n, recursive functions by an $(n + 1)$-tuple recursion that is not reducible to any lower recursion. Such functions might be called functions of minimal recursion depth $n + 1$. It may even be possible that the Paris–Harrington theorem is unprovable precisely because the patterned induction for defining $\phi(r,m,n)$ is so involved that it includes an induction on the minimal recursion depth.

Finally, we come to the last one of our heavyweights: the principle of stratification into layers by increasing complexity. This depends on a specific application of linear induction, either short or long. Historically, the first explicit occurrence of this principle is in Liouville's theory of integration in finite terms [121]. Roughly speaking, Liouville takes the class E of elementary functions $f(x)$ of ordinary calculus and decomposes it into layers $L_0, L_1, \ldots$. Here L_0 consists of algebraic functions, and each next layer is algebraically generated out of members of the previous one by adjoining a finite number of their logarithms or exponentials. The trigonometric functions are reduced to complex exponentials and their inverses to complex logarithms. Since exponentials survive differentiation and logarithms do not, Liouville is able to show that an algebraic function has an elementary integral if and only if that integral is of the form

$$f_0(x) + \sum_{j=1}^{n} c_j \log f_j(x),$$

where the c_j are constants and the $f_j(x)$ are algebraic functions.

Another instance of the stratification principle occurs in Hölder's theorem [64, 72]: The Γ-function $y = \Gamma(x)$ does not satisfy a polynomial differential equation

$$P(x, y, y', \ldots, y^{(n)}) = 0.$$

Here the layers refer to some integer concept of size, degree, or weight of the polynomial P, and the nonexistence proof runs roughly as follows. If $\Gamma(x)$ satisfies a polynomial differential equation, then there must be a simplest such equation, that is, one in the lowest layer. But the recurrence relation $\Gamma(x + 1) = x\Gamma(x)$ forces an even further simplification; thus a contradiction results and no such equation exists.

Perhaps the deepest example of the stratification principle occurs in Gödel's result [56] that if the set theory S without the generalized continuum hypothesis (G.C.H.) is consistent, then it remains so after G.C.H. is added to it as an axiom. The proof uses model theory [13]: A theory is consistent if and only if it has a model. Therefore, it is necessary to obtain a model for S + G.C.H. *within* S. A geometrical analogy might help here: to show that the hyperbolic plane geometry is consistent if the Euclidean one is, we show that with a suitable interpretation of points, lines, and operations on them the interior of a Euclidean circle is a model for the whole hyperbolic plane. Using the stratification principle, Gödel obtains a similar restriction of the class of all sets of the theory S (corresponding to the Euclidean plane) to a smaller class L (corresponding to the interior of the circle), and L turns out to be a model for S + G.C.H. within S. Call a set X constructible over the set Y if X is obtained by well-formed formulas of S in which all parameter sets (i.e., bound variables) are in Y. The Gödel hierarchy $L_0, L_1, L_2, \ldots, L_\alpha, \ldots$ is obtained by transfinite induction, starting with L_0 as the empty set, iterating the constructibility operation at nonlimit ordinals, and collecting all the previous layers at limit ordinals. The union of the layers L_α over all ordinals α is the class L of constructible sets. Within L the constructibility muzzles the indefiniteness of the power set formation

$$2^X = \text{set of } all \text{ subsets of } X$$

so strongly that here Gödel is able to *prove* the G.C.H.:

$$\text{if } \ |X| = \aleph_\alpha, \qquad then \ \ |2^X| = \aleph_{\alpha+1}.$$

Chapter Three

About This Book

Chapter 2 contained descriptions of some candidates for the "working mathematical principles," ending up with the three items of reductio ad absurdum, induction, and stratification into layers by increasing complexity. There had also been other heavyweight candidates; one of them, the principle of conjugacy, was observed quite early in the game. It was said before that this principle was first found to operate in many branches of mathematics and to concern there a variety of topics from elements to profundities. At that time it was still placed on a par, so to say, with other heavyweight principles. But later its range was seen to extend to communications and power technology, to physics and other natural sciences, and then far afield beyond science or technology. At first, the conjugacy principle appeared to be the nearest thing to a universal paradigm and even now, on a more detached view, it seems to be in a different class from all other principles, in fact to form a class by itself.

Along with those jumps in the scope of applicability or reach there went changes in the scale of projected treatment. What started as a part of an appendix in an earlier book [96], became a project for an expository paper, then for a memoir, then it split into several such projects, and finally it is winding up as a book. On account of its coverage this book has forced its author to strike an uneasy balance between incomprehensibility and non-professionalism or between gobbledygook and glibness, to say nothing of having to treat some essentially foreign subjects. Well-meaning friends and others have counseled against going out on a limb by taking up topics beyond one's competence and have recommended to the author to stay with, or close to, his subject. It was interesting to note that the counselors of caution were often more experienced but also older, though not necessarily chronologically, while the fellow enthusiasts, if not young, were generally youthful in outlook. There came also a realization that this made a sort of sense: Bypass deals with novelty, age and experience distrust it. At any rate, the belief of having something new and interesting to say tipped the scale in favor of the wide treatment rather than the narrow one and in favor of taking the chance of appearing nonprofessional.

After the present introductory part, the first aim of this book is to provide in Part 2 a variety of mathematical examples of conjugacy, together with sketches of proofs and some methodological comments. In a few places the proof outlines are amplified, either because the theorems are good but relatively unfamiliar examples of bypass (e.g., Blaschke's characterization of the ellipsoid) or because the proofs are believed to be new and have been obtained by a deliberate application of the bypass principle (Routh's theorems, the inclusion-exclusion identity). A characterization of a large class of groups in terms of three bypass axioms is also given. However, there are also present in Part 2 some lengthy statements of what is quite well known, for instance, the similarity of matrices. Such things are merely meant to illustrate and exhibit the bypass method using familiar material and to emphasize certain bypass details. A separate section is devoted to the discussion of Gödel's incompleteness theorem in the conjugacy framework, with special reference to the property of reflexness (or self-referentiality).

Part 3 contains conjugacy examples drawn from sciences other than mathematics, from technology, and from various other disciplines. The examples include the Hamilton–Jacobi equations, the methods of obtaining very low temperatures, an approach to the formalism of quantum mechanics, and a treatment of the Lorentz transformation from special relativity. Chapter 10 gives examples in communication theory and technology, principally on modulation, coding, telephony, and other information-transport problems.

The final Part 4 is admittedly speculative. Chapter 11, "Language and Meaning," concerns such matters as rhetoric, rhetorical figures, parsing of sentences and phrases, metaphors and analogies, and ends up with a hypothesis on the existence of an "inner language" of the brain. Some "models of man" and questions of human evolution are briefly considered in the conjugacy framework in the final Chapter 12.

Many of our conjugacy examples refer to transport and communication. This may be the transport of abstract structure as in many examples of the mathematical Part 2. More concrete cases of the transport of matter, energy, or information occur in Part 3. In this connection—with special reference to communication—it is suggested in Part 4 that the conjugacy principle plays an important role as the basis of symbolization. In fact, we more than hint that the most important human symbolizing communication process, language, is a case of bypass. With crude schematizing of the version (C_2) in Chapter 1, it might be said that W represents direct thought transfer from brain to brain (e.g., telepathy), S is the symbolizing correspondence between thoughts and sounds, T is the acoustic phenomenon of sound transfer, from utterance to reception, and S^{-1} is the reverse correspondence between sounds and thoughts.

For the sake of better precision it is necessary to (attempt to) explain, and unfortunately thereby to complicate, certain things about bypass right now, even at this introductory stage. We wish to discuss in more detail the bypass equation $W = STS^{-1}$, but this detailing is itself to be regarded as rough and preliminary. Final descriptions and definitions should come at the end of the book if at all, not at the beginning. Right now, let us recall the four questions of the hypothetical philosopher from the ninth preliminary example in Chapter 1:

1. What do W, S, T, S^{-1} stand for?
2. In what sense are S and S^{-1} inverses?
3. How are S, T, and S^{-1} combined into STS^{-1}?
4. What does the equality sign in $W = STS^{-1}$ assert?

There is an obscure and elusive but sometimes oddly suggestive piece of mathematical folk wisdom according to which mathematical objects correspond to nouns, while mathematical operations and transformations correspond to verbs. This offers the possibility of answering the four questions and even of answering them in two different ways. However, with any degree of linguistic sophistication, the difference between nouns and verbs is not as sharp as it might appear at first. Rather than using the old grammatical categories of nouns and verbs, we shall therefore call those two ways substantive and performative. The first is meant to refer to objects or things (somewhat as in anatomy); the other concerns producing or doing those objects or things (rather as in physiology).

For example, in elementary linear algebra W, S, T, S^{-1} may refer to mathematical objects called matrices. There is no difficulty about the points raised in questions 2–4, and the bypass equation $W = STS^{-1}$ has an obvious substantive interpretation: It states or defines the similarity of the matrices W and T. But W, S, T, S^{-1} may also stand for linear transformations of a vector space, as well as for matrices. Thus we may say that the bypass equation $W = STS^{-1}$ has a performative interpretation: It gives a necessary and sufficient condition for two matrices W and T to describe the same linear transformation in two bases.

Actually, our labored metaphor of digging shafts and tunneling provides a good example for the two interpretations. On the substantive view S, T, and S^{-1} stand for the two shafts and the connecting tunnel; $W = STS^{-1}$ merely asserts that W consists of the system of all three linked together. On the performative view, S stands for the operation of digging down, T for that of tunneling, and S^{-1} for that of digging up. We claim that the performative interpretation is both wider and more exact. Substantively, there should be no difference between S and S^{-1}; except for the accident (which we shall discuss) of being in different places, they are the same thing. But it is entirely possible that digging down calls for a technique other than that of digging up. This is a point of considerable methodological importance to conjugacy in general. In many of our bypasses S ... S^{-1} the first and the last part concern a single transformation, once directly as in S and once in reverse as in S^{-1}. Therefore, it is important to bear in mind that S and S^{-1} may require different means. Thus in telephony, the change of voice into fluctuating current may use different means and even different hardware from those required for the opposite change of fluctuating current into voice.

Returning to our performative interpretation of digging and tunneling, we ask: What is W now? This seems to be connected to question 4, and the two questions might even be merged by asking: What is "W =" now? A risky but suggestive answer would be: "W =" stands for "a way to get past the wall is by... ." Thus the complete bypass equation $W = STS^{-1}$ might then be read as "a way to get past the wall is by the operation of digging down, followed by the operation of tunneling, followed by the operation of digging up." Of course, this suggested interpretation for "W =" is tentative and imprecise. That might even be the necessary price to pay for attempting to formalize such things as escapes and novelty.

A further remark concerns another bothersome point: the entry shaft S is at one place, say at q_1, the exist shaft S^{-1} is at another, say at q_2. It might be possible to fix things up by some amended form of the bypass scheme, such as

$$W = S(q_1)TS^{-1}(q_2) \quad \text{or} \quad W(q_1,q_2) = S(q_1)T(q_1,q_2)S^{-1}(q_2). \quad (1)$$

These forms complicate notation and other matters, besides leading to trouble with respect to question 2. It might be claimed that the above two extensions of the bypass equation seem to go with the substantive interpretation, since the location parameters q_1 and q_2 answer the question Where? which is usually the first piece of information one wants to know regarding objects or things. The performative interpretation is concerned with doing or producing, that is, with activity. Here the most frequent first question concerns the control of that activity and is When? But the very syntax in $W = STS^{-1}$ answers that. Scanning is in the usual order from the left to the right and so $W = STS^{-1}$ says: Do S first, then do T, and only after that do S^{-1}.

Even a very superficial acquaintance with computing suggests that in the preceding performative interpretation the parts S, T, and S^{-1} of bypass appear as something like the instructions of a program. It is useful to observe here that computer instructions are par excellence performative; they are of the form do..., transfer..., go to. ... Since they start with the verb, and use this verb in the imperative mode, they really ought to be called commands rather than instructions. Even such a simple arithmetical instruction as $k = k + 1$ has the same imperative form and stands for "increase the value of the index k by 1" or "add 1 to the contents of location k." Further, let us recall some standard computer terminology. The physical parts of the computer, including the central processor, arithmetical units, memory, and the peripheral input/output equipment are called the hardware. In distinction to this, there is the software, which embodies the logic and/or control of operations. There appears to be no unanimous agreement on the definition of software. Some computer people apply this term to any program or programs. More often, it is used to denote a structured collection of programs, rules, data and procedures, designed not so much to solve users' specific problems as to run the computer activity as a whole. Briefly, software refers to large-scale supervision, or to master control. The preceding considerations suggest an alternative description of bypass, tentative yet radical. Bypass is a feature of natural software and especially, though not exclusively, of human software.

A self-critical spirit seems proper to end this chapter in, and so we shall list at least some possible objections to our treatment, together with brief excuses or justifications. The terminology used (bypass, level, stacking) may seem inappropriate or misleading—we prefer to use simple and standard words. In some places it may appear that there are many irrelevant side trips and a lot of unnecessary detail—since it is, after all, unfamiliar ground we tread it is hard to predict which paths and directions will prove useful and important. With respect to some topics it will perhaps be thought that the conjugacy principle has been overstretched for the sake of seeming universality—the excuse used before applies here too. An uneasy course had to be steered in this book to avoid both the sterility of rigor that explains everything and the looseness that explains nothing—no excuse is offered here. As usual, the faults of omission are perhaps worse than those of commission; two of the former appear to us particularly serious. First, there are no examples in the life sciences (biochemistry, biology, psychology; neurophysiology, immunology, anesthesia, psychiatry, medicine generally)—the only excuse is our ignorance. Second, there is nothing like a statistical approach or a treatment of conjugacy with "errors" or, more simply, consideration of bypasses where the "return" is not exact. Such a treatment could be undertaken now, but it seems too early to do so before a considerable amount of interaction with different specialists and generalists has occurred.

PART TWO

EXAMPLES IN MATHEMATICS

Chapter Four

Examples in Geometry

Ex. 4.0. THALES AND THE SCIENCE OF PROPORTION

The mathematical examples start with Thales (late seventh and early sixth century B.C.), an ancient Greek geometer who was the first mathematician individually known in history as such. In particular, we begin with one specific piece of his work, the science of proportion. This has started trigonometry, geometry, 'and algebra, though the last one indirectly, along the modern deductive lines rather than along the old Egyptian and Mesopotamian lines which are empirical. For our present purpose it is (perhaps) irrelevant that Thales might also pass for the first modern European philosopher, at least inasmuch as he postulated for the first time a natural rather than a super-natural principle: water. However little sense this makes philosophically, it certainly makes sense biologically. This may perhaps be the reason why that particular item of Thalesiana has been reported to us by the biologically, or rather, organically minded Aristotle.

Of interest here is the work of Thales on the determination of heights and distances of inaccessible objects by employing proportion and similarity. An inaccessible object is taken to be one not allowing measurement by such ordinary methods as the use of stretched rope; however, it is open to observation: It is visible. Here it may be good to remember that at a few removes the work of Thales on measuring heights of pyramids or distances between ships at sea leads to measuring distances between planets, stars, and galaxies. Accounts of Thales' work are incomplete and differ considerably among themselves; detailed references are given in [79]. In particular, we quote from that source a passage concerning the account given by Plutarch of Thales' famous measurement of the height of a pyramid in Egypt. "...the height of a pyramid is related to the length of its shadow exactly as the height of any measurable object is related to the length of its shadow at the same time of day." It is understood here that both the pyramid and the measurable object stand vertically on flat horizontal ground.

| In modern parlance the Thalesian constancy of ratios asserts the existence of the trigonometric function tan x, where x is the angle of inclination of the sun's rays to the horizontal. If the ratios are inverted by taking the length of the shadow over the height, then the reciprocal function cot x is obtained. It is of interest to observe here that an old name for the cotangent was *umbra recta* meaning "direct shadow," while the tangent was called *umbra versa* which means "inverse shadow" or "turned shadow," [27, p. 748].

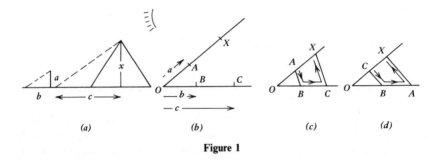

(a) (b) (c) (d)

Figure 1

Let a and b be the height of the vertical rod and the length of its shadow; similarly, let x and c be the height of the pyramid and the length of its shadow, as in Figure 1a. Disregarding the fairly easily manageable problem of measuring c, we now have three known quantities, a, b, c, and one unknown quantity, x. The determination of x depends on the proportion

$$\frac{a}{b} = \frac{x}{c},\qquad(1)$$

from which one has simply $x = ac/b$. Where and how does the bypass enter here? We should like to present the proportion (1) in the bypass form, annotating it by arrows, as

$$\left\downarrow\frac{a}{b} = \frac{x}{c}\right\uparrow \qquad(2)$$

which would presumably be read as "*a* to *b* is as *c* to *x* reversed." While correct, this might strike us as not only a contrived but even an artificial catch to force things into the shape we want. But here an allowance must be made for the arithmetization and algebraization that has taken place in the 2600 years separating us from Thales, especially in the last four centuries of that period, since Descartes. Thales reasoned geometrically, not algebraically or analytically. So, let us see how the height *x* of the inaccessible pyramid is obtained geometrically, in the ancient Greek fashion; this will suggest the geometrical essence of proportion as well as its bypass character.

We start with two straight lines intersecting in a point O as in Figure 1*b*, at some angle, the size of which does not matter. On the sloping line a point *A* is determined so that $OA = a$; similarly, on the horizontal line, two points *B* and *C* are determined so that $OB = b$ and $OC = c$. Somewhere on the sloping line lies a point *X*, such that O*X* is the height *x* of the pyramid. The fundamental observation of Thales, sometimes known as Thales' theorem, is that the proportion (1) occurs if and only if *CX* is parallel to *BA*. Hence *X* is produced by drawing the straight segment *AB*, moving over to *C*, and drawing *CX* parallel to *AB*: The point *X* is the intersection of the oblique straight line with the straight line through *C* parallel to *AB*. Now the very sequence of the three geometrical steps in Figure 1*c* justifies the apparently unnatural schema (2). The seemingly contrived arrows in (2) are seen to correspond simply and in the correct order to the three steps of the explicit geometrical program for finding the unknown *X*. It might even be said that there is here a small step toward justifying the claim made at the end of Chapter 3: Bypass is a feature of natural software.

The matter of levels between which bypasses operate was mentioned already, in Chapter 1. With reference to those levels, it is noted that the four quantities *a*, *b*, *c*, *x* are distributed in two groups of two: *a* and *x* go with the vertical heights; *b* and *c* go with the horizontal shadows. Even more explicitly, the points *A* and *X* of Figures 1*b* and 1*c* lie on the oblique straight line, while *B* and *C* are on the horizontal line. The two levels refer therefore to vertical heights and to horizontal shadows.

However, the proportion (1) and the bypass-annotated statement (2) could also be written as

$$\frac{c}{b} = \frac{x}{a} \tag{1'}$$

and with arrows as

$$\downarrow \frac{c}{b} = \frac{x}{a} \uparrow \tag{2'}$$
$$\longrightarrow$$

so that the arrows again terminate on the unknown quantity x. To us (1) and (1′) are obviously the same since they represent different ways of writing the equation $ac = bx$. Since the Greeks, at the time of Thales at least, were innocent of algebra, to them (1) and (1′) would perhaps not be the same thing. The distinction may be best explained by reference to levels: instead of vertical heights and horizontal shadows, as in (1) and (2), we have in (1′) and (2′) the completely accessible object on one level and the (partly) inaccessible pyramid on the other, as is shown in Figure 1*d*.

To someone with literary pretensions the Graeco-Egyptian pyramid of Thales might invite a comparison with the Roman-Egyptian pyramid from one of Horace's most famous odes:

> Exegi monumentum aere perennius
> Regalique situ pyramidum altius...

(I have wrought up a monument more lasting than bronze/a pyramid taller than the king's edifice...). It might be judged today that Thales was more modest than Horace, and that his monument may yet outlast Horace's. Allowing ourselves a slightly purplish patch, we might say that Thales got past the wall of inaccessibility with the novelty of his proportion bypass. This set up an important pattern for future mathematical thinking.

Of course, even if it were granted that bypass explains proportion, it is neither necessarily true nor even likely that bypass explains how or why Greeks thought about proportion and, in particular, how or why Thales was led to his discoveries. Yet, there is an intimate connection between proportion and bypass; it might even be said that proportion geometrizes bypass. This is emphasized by the three diagrams of Figure 2.

Here the first diagram illustrates geometrically the extended proportion

$$a_1:a_2:a_3:\ldots = b_1:b_2:b_3:\ldots$$

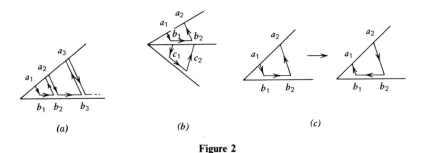

(a)   (b)   (c)

Figure 2

as well as the rule (C_4) from Chapter 1 for extension of bypasses in length. The second diagram expresses geometrically both the compound proportion

$$a_1:a_2 = b_1:b_2 = c_1:c_2$$

and the rule (C_3) from Chapter 1 for stacking bypasses in depth. Finally, the third diagram serves for inverting a proportion as in

$$a_1:a_2 = b_1:b_2 \rightarrow b_2:b_1 = a_2:a_1$$

and inverting a bypass according to rule (C_5) in Chapter 1.

An important etymological observation, which will occupy us later at some length, must be made now: The Greek word for proportion is $\alpha\nu\alpha\lambda o\gamma\iota\alpha$, which, of course, translates freely as analogy. This suggests that reasoning by analogy, and in particular solving problems and finding or describing unknown or inaccessible quantities by analogy, may sometimes turn out to depend on a proportion bypass, as in the above original problem of Thales.

Without desiring to tackle here the theologico-philosophico-linguistic subtleties of the relations between analogy and proportion, let us consider one example:

$$\text{circle is to ellipse as square is to rectangle.} \tag{3}$$

At once the questions arise: Is it a proportion, an analogy, or something partway between? How could this be made precise? How could this be used? The first question does not really interest us, except for the possibility it raises of interpolating a new term between proportion and analogy. Recalling a previous observation, we now put down some informal remarks toward answering the other two questions, but without pretending or promising any final or definite answers.

1. Like proportion, an analogy may be a relation observed between four quantities, objects, classes of objects, or classes of classes.
2. The four may be distributed in two distinct groups or on two distinct levels, two of the four on each.
3. Three of the four are assumed to be known and the purpose is to find, or describe, the fourth.
4. This purpose may (sometimes? often? always?) be achieved by using the bypass principle.

To help us validate the proposition (3) let us recall two things. The first one is that the proportion (1) may be put into words as *a* is to *b* as *x* is to *c*.

The second one is that the basic mathematical connective "ϵ" in "$u \,\epsilon\, U$" stands, according to Giuseppe Peano, who introduced this terminology in [107], for "$\epsilon\sigma\tau\iota$" which means "is." The intended interpretation is of course "is a member of": u is (a member of) U. With the obvious abbreviations, one is therefore tempted to interpret (3) as

$$C \,\epsilon\, E :: S \,\epsilon\, R, \tag{4}$$

where :: is the old symbol for proportion. That is, there are in (3) two levels: the level of individual objects and the level of classes of objects. Abstracting from mere size and position, we get the two shapes, the circle C and the square S, as individuals, whereas ellipse E and rectangle R are really classes of shapes. The relation (4) represents at least partial progress toward validating (3); it asserts that the individual C is a member of its class E in the same way in which the individual S is a member of its class R. The idea is to specify just how C is special in E in the same way in which S is special in R. It is still necessary to explain the phrase "in the same way" of the last two sentences. However, *here* we have a common mathematical situation and we know how to deal with it: by suitable transformations (projections, contractions). Let us associate with the circle C its circumscribing square, and with an ellipse E its principal circumscribing rectangle R. Then the same family of projections carries C onto all members of E and S onto all members of R.

Diagrammatically, we can put all this into the bypass form

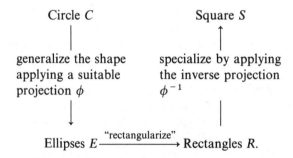

Of course, this is not a very interesting bypass because all four things C, E, R, S are well known, and we have no interest in squaring circles. It seems to be even less productive to recast (3) as

circle is to square as ellipse is to rectangle,

where the two levels of description are "rounded figures" and "rectilinear figures."

However, the main point is that the preceding remarks appear to schematize the type of mathematical thinking which employs analogy and is used in problem solving. In particular, the bypass with ellipses and circles provides us with several simple examples (Ex. 4.1–4.4) and one not so simple example (Ex. 4.8).

EX. 4.1.

Consider the following elementary problem in plane geometry. An ellipse E is inscribed into a triangle H, to show that the three straight segments joining the tangency points to the opposite vertices of H are concurrent. Project the plane P of E and H onto another plane P_1 by an orthogonal projection S, so that H projects onto some triangle H_1 while E projects as the circle C inscribed into H_1. For H_1 and C, the three straight segments are easily shown to be concurrent by an application of Ceva's theorem [46]. This easy demonstration plays the role of T. Project P_1 back onto P by the reverse projection S^{-1}, and the problem is solved. Here and in the following problems it is taken for granted that projection preserves rectilinearity.

EX. 4.2.

An ellipse E is inscribed into a convex n-gon P_n with $n = 3$, 4, or 5. Where does its center O lie? A simple intuitive reasoning shows that a general ellipse in the plane has five degrees of freedom and the tangency to a given straight line uses up one degree. Hence the center O should fill up an area for $n = 3$, an arc of a curve for $n = 4$, and be fixed for $n = 5$. Finding the locus of centers is facilitated by showing first that this locus is convex. Now one reasons as in Ex. 4.1 by projecting orthogonally so that E projects onto the circle inscribed into the projection of P_n. It turns out that for P_3 the center O lies in the triangle whose vertices are the midpoints of the sides of P_3, and for P_4 O lies on the straight segment whose endpoints are the centers of the diagonals of P_4 (Newton's theorem [42]). In each case the ellipse is uniquely determined by its center. For P_5 the center of E is fixed by Brianchon's theorem [46].

EX 4.3.

An ellipse E is inscribed into (circumscribed about) a given triangle H. When is its area maximum (minimum)? Here one works with the ratio of areas, that of the ellipse to that of the triangle, which remains unchanged by an orthogonal projection S. Let S be such that the projected triangle H_1 is equilateral. The desired extremal ellipses are obtained by applying S^{-1} to the inscribed and the circumscribed circles of H_1. As a small bonus, it has been shown that the minimizing and maximizing ellipses are homothetic, and the ratio of their areas is 4.

EX. 4.4.

In a given ellipse E, all chords are drawn which cut off a fixed fraction k of its area, $0 < k < \frac{1}{2}$. Show that the chords are tangent to a homothetic ellipse. This is done by the same projection of E onto a circle as in Exs. 4.1 and 4.3.

EX. 4.5.

Let L_1 and L_2 be two straight lines in the plane and let P, Q be two points, as shown in Figure 1. It is desired to find the shortest polygonal path PX_1X_2Q, where X_1 is a point on L_1 and X_2 is a point on L_2.

   That is, we wish to find the path of the light ray which is projected from P and reflects in L_1 and L_2 as mirrors so as to hit Q after two reflections. The well-known method of solution by images and reflections is conveniently and completely given by the bypass principle. Let S_i stand for mirror reflection

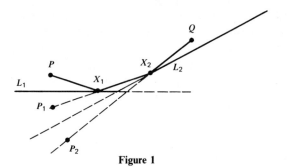

Figure 1

in the line L_i as mirror; S_i^{-1} stands for the reverse reflection from behind the mirror. The solution of Figure 1 is given by the following bypass sequence of the type (C_4) from Chapter 1:

$$S_1 : P_1 = S_1 P$$
$$S_2 : P_2 = S_2 P_1$$
$$T : X_2 \text{ is the intersection of } QP_2 \text{ and } L_2$$
$$S_2^{-1} : X_2 X_1 = S_2^{-1} X_2 P_2$$
$$S_1^{-1} : X_1 P = S_1^{-1} X_1 P_1.$$

It may be noted that the minimal path PX_1X_2Q can be found from the preceding bypass recipe without any geometrical constructions or drawing: We merely bend the paper suitably (and use a pin). Also, the bypass $S_1 S_2 T S_2^{-1} S_1^{-1}$ provides the five actual steps of the geometrical program to solve our problem, just as the proportion bypass provided the three steps of Figure 1c in Thales' problem. Thus there is here another small justification of the claim made at the end of Chapter 3 that bypass serves as a feature of natural software.

EX. 4.6. ROUTH'S THEOREMS

Let ABC be a triangle and let the straight lines AX, BY, CZ of Figure 1a be given by the ratios

$$BX/XC = l, \qquad CY/YA = m, \qquad AZ/ZB = n$$

As usual, the ratios l, m, n may be positive or negative; the positive case is

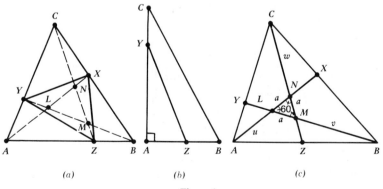

(a) (b) (c)

Figure 1

illustrated in Figure 1a. The following expressions were given without proof in 1891 by E. J. Routh [124], who needed them in network stress analysis:

$$\frac{\Delta XYZ}{\Delta ABC} = \frac{lmn + 1}{(l+1)(m+1)(n+1)}, \tag{1}$$

$$\frac{\Delta LMN}{\Delta ABC} = \frac{(lmn - 1)^2}{(lm + l + 1)(mn + m + 1)(nl + n + 1)}. \tag{2}$$

Although (1) and (2) have been obtained in the context of applied mechanics, they imply at once the important and purely geometrical theorems of Ceva (i.e., AX, BY, CZ are concurrent if and only if $lmn = 1$) and of Menelaus (i.e., X, Y, Z are collinear if and only if $lmn = -1$). Only the ratios of segment lengths and of triangle areas enter in (1) and (2). Therefore, the whole configuration may be projected orthogonally onto some particularly convenient special case. For this special case the proof of (1) and (2), or of parts of (1) and (2), may be easy; finally one projects back onto the initial configuration. Thus the bypass principle occurs here in the form: project—prove for the projection—project back.

To prove (1), start with

$$\frac{\Delta XYZ}{\Delta ABC} = 1 - \frac{\Delta AZY}{\Delta ABC} - \frac{\Delta CYX}{\Delta ABC} - \frac{\Delta BXZ}{\Delta ABC}. \tag{3}$$

By using a suitable orthogonal projection it may be assumed that the angle at A is right, as in Figure 1b. Therefore

$$\frac{\Delta AZY}{\Delta ABC} = \frac{AZ}{AB} \frac{YA}{CA} = \frac{n}{(n+1)(m+1)}$$

and analogously

$$\frac{\Delta CYX}{\Delta ABC} = \frac{m}{(m+1)(l+1)}, \qquad \frac{\Delta BXZ}{\Delta ABC} = \frac{l}{(l+1)(n+1)}$$

which yields (1) on substituting into (3).

The same technique proves (2) but to simplify the matters here, the orthogonal projection is such that ΔLMN becomes (on projection) equilateral as in Figure 1c, with the side a, say. Using the various 60° angles in Figure 1c, it is found that

$$u + a = nv, \quad v + a = lw, \quad w + a = mu, \tag{4}$$

$$u = a\frac{nl + n + 1}{lmn - 1}, \quad v = a\frac{lm + l + 1}{lmn - 1}, \quad w = a\frac{mn + m + 1}{lmn - 1}. \tag{5}$$

Since

$$\Delta ABC = \Delta LMN + \Delta ALB + \Delta BMC + \Delta CNA,$$

and the area of a triangle is half the product of two sides and the sine of the enclosed angle, one has

$$\Delta ABC = \frac{\sqrt{3}}{4}[a^2 + (a + v)u + (a + w)v + (a + u)w].$$

Therefore,

$$\frac{\Delta LMN}{\Delta ABC} = \frac{a^2}{a^2 + (a + v)u + (a + w)v + (a + u)w} = \frac{a^3}{(a + u)(a + v)(a + w) - uvw}$$

which, in view of (4) and (5), yields (2).

EX. 4.7. STEINER PORISM

Let K_1 be a circle and K_2 another circle inside K_1 in the general position (i.e., eccentric to K_1). Let C_1 be a circle tangent internally to K_1 and externally to K_2, and let $C_1, C_2, C_3, \ldots$ be a chain of circles each of which is tangent to the previous one as well as to K_1 (internally) and to K_2 (externally). It may happen that $C_1, C_2, \ldots, C_n$ is a Steiner chain: After winding round K_2 some integral number of times, the chain is such that C_n touches C_1 externally. A theorem of Steiner states that the property of being a Steiner chain depends on K_1 and K_2 alone: If it occurs for one initial position of C_1, then it occurs for every initial position [46]. A simple proof uses the bypass principle STS^{-1}, where S is an inversion that carries K_1 and K_2 onto a pair of concentric circles. The proof T is obvious for the concentric case and then S^{-1} carries the concentric circles back onto K_1 and K_2.

EX. 4.8. BLASCHKE'S THEOREM ON THE ELLIPSOID

Let K be an ovaloid, that is, a smooth strictly convex closed surface in the Euclidean space E^3. For any vector $\bar{u}$ let the equator $K(\bar{u})$ of K be the intersection of K and the circumscribing cylinder of K which projects it along $\bar{u}$. More descriptively, $K(\bar{u})$ is the shadow boundary when the surface K is illuminated by parallel rays arriving in the direction of $\bar{u}$ from infinity. Blaschke's theorem is: If all equators of K are plane curves, then K is an ellipsoid. Following Blaschke [18], assume first only that $K(\bar{u})$ is plane for all $\bar{u}$ parallel to P, and take the plane P to be horizontal. Let p and q be the two points of support of K by horizontal planes. Since the property assumed is unchanged by affine transformations, it may be supposed without loss of generality that pq is perpendicular to P. Incidentally, the phrase "without loss of generality" is seen, on closer examination, to be a quiet application of the bypass principle, as is quite often the case in mathematical work. Here, for instance, one applies first an affine transformation S under which the image of pq is perpendicular to the image of P. This affinity S preserves all the necessary properties, and the desired conclusion is proved for the convenient "special case." The proof for this special case plays the role of T, the middle part of the bypass STS^{-1}. Finally the reverse affinity S^{-1} is applied, and the conclusion is shown to hold in general.

Let P_1 and P_2 be two planes parallel to P and cutting K, and let $\bar{u}_1$ be also parallel to P. By considering parallel tangents to $P_1 \cap K$ and to $P_2 \cap K$ at points where $K(\bar{u}_1)$ cuts these two sections, it is shown that all sections of K by horizontal planes are similar, and that every equator $K(\bar{u})$, where u is parallel to P, has pq as its symmetry axis. Suppose now that every equator of K is plane. Then the following essential observation follows from the preceding: Every equator $K(\bar{u})$ of K is affine-symmetric with an affine symmetry axis pq, where p and q may be taken to be any pair of points on $K(\bar{u})$ with parallel tangents.

Next comes the use of the bypass principle. Fix an arbitrary equator $K(\bar{u})$ and let A, A_n be its two affine symmetry axes, such that the product of affine reflections in those axes is an affinity ϕ of period 2^n. Then, as Blaschke shows, there is a conjugating affinity ψ such that $\psi\phi\psi^{-1}$ is a rotation of $\psi K(\bar{u})$ by the angle $2\pi/2^n$. Since the latter can be made arbitrarily small, it follows that $\psi K(\bar{u})$ is a circle and so $K(\bar{u})$ is an ellipse. Thus every equator is an ellipse and from this it follows that K is an ellipsoid.

EX. 4.9. SYNTHETIC AND ANALYTIC GEOMETRY

It is well known that already the ancient Greek geometers had developed a fairly complete theory of conic sections. This culminated in the work of Apollonius of Perga and proceeded by the synthetic method. In that approach there is present a certain bypass, so obvious that it is apt to be missed altogether. Apollonius (let us say) is interested in the properties of a plane curve, the ellipse. To obtain them, he passes from the two-dimensional plane to the three-dimensional space, works in that space by intersecting surfaces in it and doing similar clever three-dimensional things, and then returns to the plane and the ellipse. Thus, in effect, he achieves his aim by the conjugacy $W = STS^{-1}$ where S is the passage from the Euclidean plane E^2 to the Euclidean space E^3, and T is the synthetic work in E^3.

Descartes, too, is interested in the properties of the ellipse, but he never moves out of the plane into the space and manages instead to abolish the wall of the Apollonian bypass by his analytic geometry. Of course, he accomplishes this by applying another bypass $W = S_1 T_1 S_1^{-1}$. Here S_1 is the correspondence between points, lines, and curves on the one hand, and numbers, variables, and relations between them on the other. T_1 is the analytic work: doing clever things with numbers, variables, and equations, and S_1^{-1} is the return to the geometric entities. It might be argued that the Cartesian analytic bypass is profounder than the Apollonian synthetic one because the level difference between equations and loci far exceeds the level difference between plane and space.

The Descartes correspondence S_1 starts with the usual coordinatization fixing points by their ordinary Cartesian coordinates. The modern counterpart of this process is prominent in the so-called geometric algebra. In this discipline various axiomatically given geometries are coordinatized by n-tuples of elements from certain algebraic structures (such as division rings, skew fields, and fields) and conversely, an algebraic structure is used to define a geometry. Even a superficial look at the principal book of the subject, Artin's *Geometric Algebra* [6], shows a heavy and varied use of conjugacy.

We note, incidentally, the presence of the Apollonian bypass STS^{-1} in connection with Desargues' theorem [6]. In the axiomatics of *plane* projective geometry this theorem is taken as an axiom, and it is well known to be equivalent to the associative law in the domain of coordinatization. However, if the two-dimensional plane can be embedded in the three-dimensional space, then the Apollonian bypass STS^{-1}, with a suitable T, yields a *proof* of Desargues' theorem.

A simple example of what has been called the Apollonian bypass (plane to space—work in space—from space back to plane) occurs in the following geometrical problem. Three unit circles in the plane pass through a point P

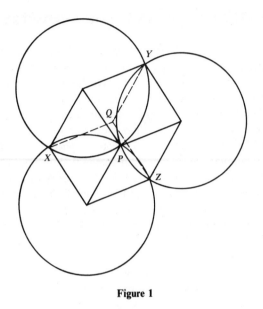

Figure 1

as is shown in Figure 1. Construct the circle C through the other inter-section points X, Y, Z. Let the center of each circle be joined by a straight segment to the three points in which the other two circles cut it, as in Figure 1. One then gets six rhombi of unit sides; in particular, C is also a unit circle centered at some point Q. However, it is much easier to see the whole situation in space even though the "plane formulation" may seduce us into "plane thinking." Now the six rhombi form a perspective drawing of a unit cube and so the center Q of C is the vertex of the cube antipodal to P.

In conclusion, it must be added that certain forms of the Apollonian bypass may be detected outside the domain of geometry. They are all comprehended in the schematic description $n/m/n$: Start with a situation in n dimensions—pass, for convenience, to a situation in m dimensions and work there—return to n dimensions. Under this code the Apollonian bypass proper becomes 2/3/2. A simple and very well known instance of 1/2/1 occurs in evaluating the important integral

$$I = \int_{-\infty}^{\infty} e^{-x^2} dx.$$

The substitution $x^2 = u$ shows that

$$I = 2 \int_0^\infty e^{-x^2} dx = \int_0^\infty u^{\frac{1}{2}-1} e^{-u} du = \Gamma(\tfrac{1}{2}).$$

To evaluate the "one-dimensional" integral I, we multiply I by itself and represent the square as a "two-dimensional" integral:

$$I^2 = \int_{-\infty}^\infty \int_{-\infty}^\infty e^{-x^2-y^2} dx\, dy.$$

Polar coordinates $x = r \cos \theta$, $y = r \sin \theta$ are introduced to yield

$$I^2 = \int_0^{2\pi} \int_0^\infty e^{-r^2} r\, dr\, d\theta$$

and the trivial θ-integration is carried out; this "collapses" the two-dimensional situation back to one dimension

$$I^2 = 2\pi \int_0^\infty e^{-r^2} r\, dr,$$

The integral is evaluated to be $\frac{1}{2}$ by putting $r^2 = v$, and thus $I = \sqrt{\pi}$.

An intriguing though questionable instance of a 3/4/3 form of the Apollonian bypass might be used to explain a seemingly strange situation in photography and portrait painting: A face will either confront the beholder always straight on or never straight on, no matter what the angle of viewing is. A possible explanation may hinge on the fact that 2 three-dimensional subspaces (the "space" of the image and the "space" of the beholder) of a four-dimensional space (the "proper space" including both) may share a two-dimensional plane only (the "plane" of the painting or the photograph). Or they may coincide. The direction of viewing, both for the beholder and for the image, is at right angles to those 2 three-dimensional spaces. Thus, in the former case the viewing is never straight on, and in the latter it is always straight on.

This sort of thing has been noted from time to time and perhaps never more strikingly than in the Latin treatise "De Visione Dei" [35], written in 1453 by Nicolaus Cusanus. We quote from the preface:

> If I am trying to elevate you to divine matters in a way suited to men, it ought to be done by means of some sort of analogy. I have, however, found no likeness of one who sees all that better suits our purpose: it is the fact that a face painted with refined skill has the property of seeming to face everything. Although many

of these, excellently painted, turn up—the one of the archer in the square of Nürnberg, for example, and in a precious painting by the great painter Roger kept in the town hall of Brussells, and in my chapel of Veronica at Koblenz, and the one at Brixen of an angel holding the arms of the church, and numerous others all over the place—nonetheless, so that you are not short of one in the experiment I have come up with, which needs such a picture, by your leave I am sending a painting containing a figure of Him who sees all, and I call it an icon of God. Put it up in some place, on the north wall for instance, and let all you brethren be standing round it, at equal distances. Now look at it, and let each of you discover that, from wherever he looks at it, he is watched as if alone by it. And it will seem to the brother who was placed to the east that the face stares eastward, and the same for those placed in the south and west ... even if (a brother) walks from west to east, while fixing his gaze on the icon, he will discover that the icon's visage will stare at him all along, nor desert him though he turn back from east to west, and he will wonder how it moves motionlessly.

EX. 4.1. DIFFERENTIABLE MANIFOLDS

The central unifying concept in modern geometry is that of a differentiable manifold. It arose gradually from various sources: dynamical systems, Riemann surfaces, algebraic varieties, Lie groups, and so on. To examine what is behind it, let us start with the celebrated Erlanger program, the name given to Felix Klein's inaugural address *"Vergleichende Betrachtungen über neuere geometrische Forschungen"* (Comparative Considerations on Newer Geometrical Researches) at the University of Erlangen, 1872. According to Klein, a geometry is the study of those properties of a space S, which remain invariant under a group G of transformations of S onto itself.

For instance, S may be the familiar Euclidean space E^2 or E^3, that is, the set R^2 or R^3 with the Euclidean metric structure; G may be the group of rigid motions, reflections included. This leads to the usual plane or solid geometry, with its study of congruence and similarity. Or the same set R^2 or R^3 may be given the affine rather than the Euclidean structure; G is then the group of affinities, and we get the affine geometry (of A^2 or A^3). Or R^2 may be augmented by the addition of certain ideal elements (points or lines at infinity), and the group G may be that of inversions or projective transformations; we then get the geometry of the inversive or of the projective plane.

For certain technical reasons [19, p. 26], Klein's definition may be loosened somewhat: Geometry is the study of S when the invariance under the group G is replaced by equivalence under an equivalence relation $X \sim Y$ defined between the subsets of S. For instance, when S is arbitrary and $X \sim Y$ means that X and Y are homeomorphic, the resulting geometry is called the topology of S. In Klein's day, this (extended) definition embraced all then known geometries, but it also turned out to apply to certain things that came after. In 1912, Klein identified the then recent theory of special relativity as the study of those properties of the four-dimensional space-time continuum, which are invariant under the Lorentz group of collineations.

Yet, Klein's prescription is far too restrictive in its insistence on group invariance in the large, over the whole space S. Let S be a plane or a sphere, and let it be pushed slightly out of shape. Even though the perturbation is smooth and only local, the perturbed surface is no longer admissible for a Kleinian geometry, although clearly it has kept most of its deeper geometrical structure. This leads to the idea of replacing global geometrical properties by local ones. Since the most familiar geometrical object is the Euclidean space E^n, we arrive at the concept of a manifold: a sufficiently nice topological space that is locally Euclidean. For "sufficiently nice" we take a Hausdorff space with a countable basis. "Locally Euclidean" means that every point has a neighborhood homeomorphic to an open subset of some Euclidean space. The additional structure, which will be imposed on a manifold, together with Brouwer's theorem on the invariance of dimension, forces all the local Euclidean dimensions to be the same number n, called the dimension of the manifold. Occasionally, the further condition of connectedness may be added to the definition of a manifold.

To be useful, the manifolds ought to be smooth since we want to apply to them the usual machinery of calculus. Therefore, it is necessary to impose and explain the differentiability structure. From our point of view it is also necessary to explain the role of bypass. It turns out that these two, differentiability and bypass, are related. Consider the example of an

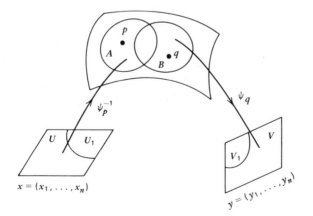

Figure 1

n-dimensional manifold M. Let p and q be the two points of M, shown in Figure 1. Let A and B be their neighborhoods corresponding to the open subsets U and V of the Euclidean space E^n under the homeomorphisms ψ_p and ψ_q. Suppose that $A \cap B$ is not empty and consider

$$\psi_p^{-1} R_{A \cap B} \psi_q; \tag{1}$$

here $R_{A \cap B}$ is the restriction to $A \cap B$, and the multiplication operation is composition. Then it follows that there are open subsets U_1 and V_1 of U and V, such that the bypass (1) defines a function with the domain U_1 and the range V_1. It must be observed that (1) is not a simple bypass of the form STS^{-1} because the map ψ^{-1} is evaluated at p and its inverse ψ at q. This is a clear example of the extended bypass schema (1) from the introduction (Chapter 3). With this understanding (1) may be called a bypass function; its domain and range are in the Euclidean space E^n and so the usual machinery of calculus is available. Obviously (1) is a continuous function but more stringent smoothness requirements may be placed on it: being in the class C^r (r times continuously differentiable), or in the class C^∞, or in the class A (real-analytic). Finally the manifold M is smooth of class C^r, or of class C^∞, or of class A, if there are sufficiently many coordinate neighborhoods and if every bypass function (1) has the required smoothness (C^r, C^∞, or A). Details are given in any text on the subject, for example, Reference 8 or 22.

An important final remark about this example concerns the use of bypass in general. As was seen already on some examples, and will be seen on many more, bypass is often used for transport purposes. In other cases we transport objects, matter, energy, information, or credit. Here we have a mathematically essential instance in which bypass serves to transport structure.

Chapter Five

Examples in Analysis

EX. 5.1. POINCARÉ'S FORMULA AND THE INCLUSION–EXCLUSION IDENTITY

Although this example really belongs to probability and combinatorics, it is given here on account of its analytical form. Let $E_1, E_2, \ldots, E_n$ be n independent elementary events, then a well-known formula asserts that

$$\text{Prob(at least one } E_i \text{ occurs)} = 1 - \prod_1^n [1 - \text{Prob}(E_i)]. \tag{1}$$

The complementation function $1 - x$, or Compl x, is its own inverse: Compl $x = \text{Compl}^{-1} x$, because $y = 1 - x$ implies $x = 1 - y$, and so (1) shows that the desired probability is obtained from the component probabilities $P(E_1), P(E_2), \ldots, P(E_n)$ by the simple bypass recipe:

$$\text{Compl Multiply Compl}^{-1}.$$

Another form of (1) is known as the Poincaré formula

$$\text{Prob (not } E_1 \text{ and } \ldots \text{ and not } E_n) =$$
$$1 - \sum_1^n \text{Prob}(E_i) + \sum\sum_{1 \leqslant i < j \leqslant n} \text{Prob}(E_i \text{ and } E_j) - \sum\sum\sum + \ldots. \tag{2}$$

77

A more general identity is

$$\prod_1^n (1 - tX_i) = 1 - t\sum_1 + t^2\sum_2 - \ldots + (-1)^n t^n \sum_n, \tag{3}$$

where t is the Eulerian parsing parameter, $X_1, \ldots, X_n$ are arbitrary symbols, and

$$\sum_k = \sum\sum_{1 \leqslant i_1 < \ldots < i_k \leqslant n} \ldots \sum X_{i_1} \ldots X_{i_k};$$

Here the summation extends over all strings $i_1, \ldots, i_k$ of integers satisfying $1 \leqslant i_1 < \ldots < i_k \leqslant n$. Let the undefined symbols $X_1, \ldots, X_n$ be interpreted with a view to a classification of some basic set S with respect to n properties simultaneously:

$$X_i S = \{s : s \in S, s \text{ has the } i\text{th property}\},$$

$$(1 - X_i)S = \{s : s \in S, s \text{ does not have the } i\text{th property}\},$$

$$X_i X_j S = \{s : s \in S, s \text{ has the } i\text{th and the } j\text{th property}\},$$

$$(1 - X_i)X_j S = \{s : s \in S, s \text{ has the } j\text{th but not the } i\text{th property}\}$$

and so on. Applying both sides of (3) to S, we have

$$\left\{\prod_1^n (1 - tX_i)\right\}S = S_0 - S_1 t + S_2 t^2 - \ldots + (-1)^n S_n t^n, \tag{4}$$

where the sets S_k arise from *inclusive* classification: S_k consists of those elements of S that have at least k properties, though possibly more. There is also the *exclusive* classification where the interest is in sets P_k consisting of those elements of S with exactly k properties. To connect P_k with (3), let us recall the standard Boolean terminology: An atom is defined as

$$\bar{X}_1 \bar{X}_2 \ldots \bar{X}_n S,$$

where each $\bar{X}_i$ is either X_i or $1 - X_i$. There are 2^n atoms, and if exactly k of the $\bar{X}_i$'s are X_i, the atom is k-tuple. Now P_k is just the union of all the $\binom{n}{k}$ k-tuple atoms. Since

$$1 - tX_i = 1 - X_i + (1 - t)X_i,$$

(3) may be written as

$$\left\{\prod_{1}^{n}[1 - X_i + (1-t)X_i]\right\}S = P_0 + P_1(1-t) + P_2(1-t)^2 + \ldots + P_n(1-t)^n. \quad (5)$$

From (4) and (5) it is seen that the generating functions of S_k and of P_k are the same, but one is in the powers of $-t$ while the other is in powers of $1 - t$. With $u = 1 - t$, this yields

$$\sum_{k=0}^{n} S_k(u-1)^k = \sum_{k=0}^{n} P_k u^k;$$

equating the coefficients of u^k leads to the basic inclusion–exclusion identity

$$P_k = S_k - \binom{k+1}{k}S_{k+1} + \binom{k+2}{k}S_{k+2} - \ldots + (-1)^{n-k}\binom{n}{k}S_n.$$

Among the many other approaches to inclusion–exclusion [120, 123], ours has the advantage of generalizing easily; an application to a multiple clustering problem is given in Reference 98.

EX. 5.2. FAA DI BRUNO'S FORMULA

It is desired here to obtain an explicit formula for the nth derivative D_x^n with respect to x of the composite function $f(g(x))$. Elementary chain rule shows that

$$D_x^1 f(g(x)) = f'g', \qquad D_x^2 f(g(x)) = f'g'' + f''g'^2,$$
$$D_x^3 f(g(x)) = f'g''' + f''(3g'g'') + f'''g'^3,$$
$$D_x^4 f(g(x)) = f'g^{(4)} + f''(4g'g''' + 3g''^2) + f'''(6g'^2g'') + f^{(4)}g'^4,$$

and so on, where $f^{(k)}$ is the kth derivative of $f(g)$ with respect to g. Suppressing the considerable combinatorial significance of the numerical coefficients [120], we show by simple induction that

$$D_x^n f(g(x)) = \sum_{k=1}^{n} f^{(k)} A_{nk}(g',g'', \ldots, g^{(n)}), \quad (1)$$

where A_{nk} does not depend on f. This independence is exploited by choosing $f(u) = \exp au$; also, to be able to apply Taylor's theorem, we take $g(x)$ to be analytic. With these choices (1) becomes

$$e^{-ag(x)}D_x^n e^{ag(x)} = \sum_{k=1}^{n} a^k A_{nk} \tag{2}$$

and the bypass schema is clearly displayed on the left-hand side. Multiply (2) by $t^n/n!$ and sum over n; using the Taylor formula one gets

$$e^{a[g(x+t)-g(x)]} = 1 + \sum_{n=1}^{\infty} \sum_{k=1}^{\infty} a^k t^n A_{nk}/n!$$

which serves as the generating function for A_{nk}. By expanding in powers of t we have

$$\prod_{j=1}^{\infty} \exp[at^j g^{(j)}/j!] = 1 + \sum_{n=1}^{\infty} \sum_{k=1}^{n} a^k t^n A_{nk}/n!,$$

and, expanding each factor by the exponential series, we obtain

$$\prod_{j=1}^{\infty} \sum_{m_j=0}^{\infty} a^{m_j} t^{jm_j} [g^{(j)}/j!]^{m_j}/m_j! = 1 + \sum_{n=1}^{\infty} \sum_{k=1}^{n} a^k t^n A_{nk}/n!.$$

Now, equating the coefficients of $a^k t^n$, one gets

$$A_{nk} = n! \sum_{m_1} \sum_{m_2} \cdots \sum_{m_s} (g'/1!)^{m_1}(g''/2!)^{m_2} \cdots (g^{(s)}/s!)^{m_s}/m_1! \cdots m_s! \tag{3}$$

where the summation is over all postive integers $m_1, m_2, \ldots, m_s$ such that $m_1 + m_2 + \ldots + m_s = k$ and $m_1 + 2m_2 + \ldots + sm_s = n$. Faa di Bruno's formula results from substituting for A_{nk} from (3) into (1).

EX. 5.3. CONFORMAL MAPPING AND BOUNDARY VALUE PROBLEMS

The conformal mapping property of the transformation $z = F(w)$, with F analytic and $F' \neq 0$, can be used to solve certain boundary value problems by means of a straightforward bypass. Suppose that the problem is as follows: Given a plane region R, find a function $\phi(x,y)$ defined on R, such that

$$\phi_{xx} + \phi_{yy} = 0, \qquad (x,y) \in R,$$

and on the boundary ∂R of R, ϕ has prescribed value. Or, instead of ϕ, the normal derivative $\partial \phi/\partial n$ is prescribed on ∂R. Or, again, ϕ is prescribed on some parts of ∂R and $\partial \phi/\partial n$ on others.

The basis of the technique is the fact that if

$$z = x + iy, \qquad w = u + iv, \qquad z = F(w), \qquad F \text{ analytic and } F' \neq 0,$$

then under the transformation

$$S : (x,y) \to (x(u,v), y(u,v)),$$

the region R in the z-plane goes over into a region R_1 in the w-plane and

$$\phi_{xx} + \phi_{yy} = (\phi_{uu} + \phi_{vv})|w'|^2$$

so that

$$\phi_{xx} + \phi_{yy} = 0 \text{ for } (x,y) \in R \leftrightarrow \phi_{uu} + \phi_{vv} = 0 \text{ for } (u,v) \in R_1$$

with similar preservation of boundary behavior. Accordingly, there is the bypass

$$
\begin{array}{ccc}
\text{boundary value problem in } (x,y) & & (x,y) \text{ solution} \\
F \downarrow & & F^{-1} \uparrow \\
\text{boundary value problem in } (u,v) & \xrightarrow{\text{solve}} & (u,v) \text{ solution.}
\end{array}
$$

This bypass is often arranged so as to be "trivializing," that is, to make the middle part trivial or at least very simple, as in the geometrical problems on the ellipse or in the example involving the Hamilton–Jacobi theory. The schema (C_4) of applying several layers in depth shows itself quite clearly here; it may be hard to trivialize the boundary value problem in one swoop, but it is easy to do so on installments and hence to obtain successive complexity reductions.

For example, consider the steady-state temperature distribution problem for a half-strip [30]: to find $\tau = \tau(x,y)$ such that

$$\tau_{xx} + \tau_{yy} = 0, \qquad -\pi/2 < x < \pi/2, \qquad y > 0,$$
$$\tau(-\pi/2, y) = \tau(\pi/2, y) = 0, \qquad y > 0,$$
$$\tau(x, 0) = 1, \qquad -\pi/2 < x < \pi/2.$$

Note that $\tau(\pm\pi/2, 0)$ is undefined and that τ is to be bounded in the whole half-strip.

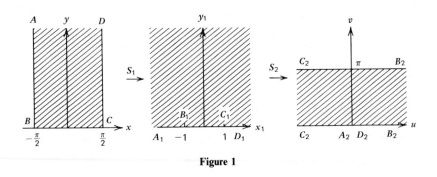

Figure 1

We apply in succession two conformal mappings, shown in Figure 1. The first one maps the original vertical half-strip onto the upper half-plane, and the second one sends this half-plane onto a full horizontal strip, with the resulting "disentanglement" of boundary values. The transformations are as follows:

$$S_1 : z_1 = \sin z, \qquad S_2 : w = \log \frac{z_1 - 1}{z_1 + 1}.$$

The middle part T of the bypass is the very easy problem associated with Figure 1c: to find the harmonic function $\tau(u,v)$ for the strip shown, such that $\tau(u,0) = 0$ and $\tau(u,\pi) = 1$; the solution is clearly $\tau(u,v) = v/\pi$. Now the bypass

$$S_1 S_2 T S_2^{-1} S_1^{-1}$$

gives us the solution of the original problem

$$\tau(x,y) = \frac{2}{\pi} \arctan(\cos x / \sinh y).$$

EX. 5.4. SERIES SUMMATION

An identity similar to (2) in Ex. 5.2 will sometimes synthesize the right series coefficients and so help in summation. For instance,

$$x^{-b} \left(x \frac{d}{dx} \right)^k x^b x^n = (n + b)^k x^n;$$

letting $\Sigma a_n x^n = f(x)$ be a power series, we have

$$\sum a_n (n+b)^k x^n = x^{-b} \left[x \frac{d}{dx} \right]^k [x^b f(x)],$$

$$\sum a_n (n+b)^{-k} x^n = x^{-b} \left(\int \cdots \frac{dx}{x} \right)^k [x^b f(x)].$$

A more serious application arises from telescoping cancellation

$$(a_1 - a_0) + (a_2 - a_1) = \ldots + (a_n - a_{n-1}) = a_n - a_0$$

in which the bypass schema appears in the form: For ease of summation express the given series as

Summation (of) Term(s which are) Summation^{-1}

if Summation^{-1} is interpreted as subtraction or, better, differencing. There is an analogous expression for products; both have numerous applications to evaluating finite and infinite series and products. Simple examples occur in evaluating

$$S = \sum_{n=1}^{\infty} \left(a_n / \prod_{j=1}^{n} (1 + a_j) \right) \quad \text{and} \quad P = \prod_{n=2}^{\infty} (1 - n^{-2}),$$

where it is assumed that

$$a_n \neq -1 \quad \text{and} \quad \lim_{n \to \infty} \prod_{j=1}^{n} (1 + a_j) = \infty.$$

We find first by telescoping cancellation that

$$S_N = \sum_{n=1}^{N} \left(a_n / \prod_{j=1}^{n} (1 + a_j) \right) = \left(1 - \frac{1}{1 + a_1} \right) + \left(\frac{1}{1 + a_1} - \frac{1}{(1 + a_1)(1 + a_2)} \right)$$

$$+ \ldots + \left(\frac{1}{(1 + a_1)\ldots(1 + a_{N-1})} - \frac{1}{(1 + a_1)\ldots(1 + a_N)} \right) = 1 - \left(\prod_{j=1}^{N} (1 + a_j) \right)^{-1}$$

and

$$P_N = \prod_{n=2}^{N} (1 - n^{-2}) = \frac{1 \cdot 3}{2 \cdot 2} \cdot \frac{2 \cdot 4}{3 \cdot 3} \cdot \ldots \cdot \frac{(N-1)(N+1)}{N^2} = \frac{1}{2} \frac{N+1}{N}.$$

Hence

$$S = 1 \quad \text{and} \quad P = 1/2.$$

A particular use may be mentioned in connection with the digamma function

$$\psi(x) = \sum_{n=1}^{\infty} \frac{x}{n(n+x)} - \gamma = \Gamma'(x+1)/\Gamma(x+1),$$

where γ is Euler's constant. This function has the first difference

$$\psi(x+1) - \psi(x) = \frac{1}{1+x},$$

and it can be used to express in closed form convergent series of the type

$$\sum_{n=1}^{\infty} \frac{P(n)}{Q(n)},$$

where P and Q are polynomials. For instance,

$$\sum_{1}^{\infty} \frac{1}{(n+a)(n+b)} = \frac{\psi(b) - \psi(a)}{b-a}$$

and

$$\sum_{1}^{\infty} \frac{1}{(n+a)^2} = \psi'(a).$$

Generally, $\psi(x)$ plays the same role in summation and differencing as logarithm does in integration and differentiation.

EX. 5.5.   ARITHMETIC, GEOMETRIC, AND HARMONIC MEANS

Let $x_1, \ldots, x_n$ be n positive numbers, and let $A(x_1, \ldots, x_n) = A(x)$ and $G(x_1, \ldots, x_n) = G(x)$ stand for their arithmetic and geometric means. A very well known inequality states that

$$G(x) \leqslant A(x), \tag{1}$$

and the inequality is strict unless the x_i are equal. Let $R(x)$ stand for the reciprocal function: $R(x) = 1/x$. Then $R^{-1}(x) = R(x)$, and we have the bypass relations

$$G(x) = RGR^{-1}(x), \qquad H(x) = RAR^{-1}(x) \tag{2}$$

of which the second one defines the harmonic mean as the reciprocal of the arithmetic mean of the reciprocals. Hence (1) extends to

$$H(x) \leqslant G(x) \leqslant A(x) \tag{3}$$

with the same condition on equality as in (1). Suppose next that it is desired to extend the various means and the arithmetic-geometric-harmonic inequality (3) to the continuous domain, that is, from a discrete set $x_1, \ldots, x_n$ of positive numbers to a positive function $f(x)$, $a \leqslant x \leqslant b$. No difficulty arises with the arithmetic mean $A(f)$, and by the definition of an integral as a limit of sums

$$A(f) = \frac{1}{b-a} \int_a^b f(x)\,dx.$$

But what about $G(f)$ and $H(f)$: How are they to be defined? Let $E(x)$ stand for the exponential function, then

$$G(x_1, \ldots, x_n) = EA(\log x_1, \ldots, \log x_n).$$

That is, in the explicit form of the conjugacy principle,

$$G(x) = EAE^{-1}(x).$$

Thus the geometric mean is conjugate to the arithmetic mean under the exponential E, and by (2) the harmonic mean is conjugate to it under the reciprocal R. Hence we define

$$G(f) = \exp\left[\frac{1}{b-a} \int_a^b \log f(x)\,dx\right],$$

$$H(f) = (b-a)\Big/ \int_a^b \frac{dx}{f(x)},$$

and we have, as in (3),

$$H(f) \leqslant G(f) \leqslant A(f).$$

For instance, applying the above with $a = 0$, $b = 1/N$, and $f(x) = (1 + ux)^u$, we get for N large

$$1 - \frac{u^2}{2N} + O(N^{-2}) \leqslant \frac{(1 + u/N)^N}{e^u} \leqslant 1 - \frac{u^2}{2N} + O(N^{-2}).$$

Many further examples of this kind will be found in Reference [111].

EX. 5.6. THE EXPONENTIAL

The powers x^a of a general number x are simply defined when a is integral or rational. For the general exponent a, the exponential function is used: $x^a = \exp(a \log x)$, that is,

$$x^a = \exp(a \exp^{-1} x).$$

In words, the (difficult) operation of raising to a power is conjugate under the exponential function to the (easy) operation of multiplication. We can even define the exponential as the function $\phi(x)$ that performs the desired conjugation

$$x^a = \phi(a\phi^{-1}(x)).$$

With $a\phi^{-1}(x) = t$, the preceding equation becomes

$$\phi(t) = \phi^a(t/a). \tag{1}$$

Setting $t = 0$ and $a = 2$, we find $\phi(0) = \phi^2(0)$ so that $\phi(0) = 0$ or $\phi(0) = 1$; the former implies under minimal smoothness that $\phi(t) = 0$ for all t. Suppose now that not only is $\phi(0) = 1$, but that for small t we have further

$$\phi(t) = 1 + Kt + O(t^\alpha), \qquad \alpha > 1. \tag{2}$$

Then (1) implies that

$$\phi(t) = (1 + Kt/a + O(t^\alpha)/a^\alpha)^a$$

and by taking the limit as $a \to \infty$

$$\phi(t) = e^{Kt}, \qquad K \text{ arbitrary.} \tag{3}$$

The theory of semigroups and their infinitesimal generators follows the same pattern. For a semigroup T_t of operators (1) follows from the semigroup property $T_{t+s} = T_t T_s$. In analogy to (2) a sufficient smoothness condition is assumed to hold near the identity $I = T_0$. The result (3) is then the Hille–Yosida theorem [71]: Under suitable conditions the semigroup T_t is representable in the exponential form

$$T_t = e^{tK}$$

where K is the infinitesimal generator of the semigroup and

$$K = \lim_{t \to +0} (T_t - I)/t.$$

EX. 5.7. ITERATION

In analysis the conjugation principle comes into its own in connection with iteration. Let $f(x)$ be a function and $f_n(x)$ its nth iterate. For a few very special functions, such as

$$x + k, \qquad cx, \qquad ax + b, \qquad x^2,$$

the nth iterate can be simply written down as an explicit function of x and n. In general, this is a hard problem, even for functions as simple as the quadratic $f(x) = ax^2 + bx + c$. As was observed by S. M. Ulam (in a personal communication), our inability to iterate nonlinearities hampers us very considerably in tackling nonlinear problems; this is obvious when, e.g., a nonlinear estimate is to be iterated in connection with the method of successive approximations, and we work without assuming the Lipschitz condition [31].

It was noticed already by Abel [1] that the problem of explicit iteration may be attacked by means of conjugacy. If $\phi(x)$ is a solution of Abel's functional equation

$$\phi(x) + 1 = \phi(f(x)),$$

so that $f(x) = \phi^{-1}\tau\phi$, with $\tau(x)$ being the translation $x + 1$, then

$$f_n(x) = \phi^{-1}\tau_n\phi(x) = \phi^{-1}(n + \phi(x)).$$

That is, Abel conjugates the "difficult" function $f(x)$ to the "easy" function $x + 1$ by means of ϕ^{-1}. Abel's work [1] appeared posthumously in 1881. Ten years earlier Schröder [126] conjugated f to the multiplication:

$$c\psi(x) = \psi(f(x)),$$

and ψ is called the Schröder function for f. However, the originator of the conjugacy method appears to have been Charles Babbage, the inventor of automatic computers. Working in 1820, or before, on the solutions of the iteration equation

$$f_n(x) = x,$$

he observed in [44] that if $f(x)$ is a solution, then $\phi^{-1}(f(\phi(x)))$ is also a solution, ϕ being more or less arbitrary. Babbage's work was commented on and continued by Ritt [122]; a recent reference is [44, Ch.4].

The conjugacy method in iteration really accomplishes a great deal more than mere explicit expression for the nth iterate $f_n(x)$ of $f(x)$. Namely, it provides a tool to attack one of the basic mathematical problems, the extension problem. This is: how to extend to a wider domain a function or an operation originally given only for a narrower domain. A classical case is the extension to all real or complex values something originally given only for integers (e.g., the factorials $n!$ or the nth derivative of a function). An example is: how to find the iterates of the Peano successor function $s(x) = x + 1$, at $x = 0$. Of course, we know that $s_n(0) = n$ is the nth iterate, and thus $s_a(0) = a$ is the ath iterate. But the problem becomes less trivial if $s_n(0)$ is taken as the von Neumann integer n [80], defined by

$$0 = \varnothing, 1 = \{\varnothing\}, \dots, n + 1 = \text{union set of } n.$$

Any number of examples of iteration of fractional, or real, order may be given. For example, we may wish to produce a solution of the functional equation

$$f(f(x)) = 4x(1 + x). \tag{1}$$

If $g(x)$ is the right-hand side, then it is observed that $g(x)$ performs the duplication for $\sinh^2 x$:

$$\sinh^2 2x = g(\sinh^2 x).$$

That is, with $\phi(x) = \sinh^2 x$ and with the simple duplication function $d(x) = 2x$, we have

$$\phi(d(x)) = g(\phi(x))$$

and so $g = \phi d \phi^{-1}$: g is conjugate to d under ϕ. Since $f(x)$ is a "half-order" iterate of $g(x)$, we obtain

$$g_n(x) = \phi d_n \phi^{-1}(x)$$

as the general nth order iterate, and so, explicitly,

$$f(x) = \sinh^2(\sqrt{2} \text{ arg sinh } \sqrt{x})$$

is a solution of (1). We do not go here into the knotty problem of uniqueness, that is, of sufficient smoothness conditions to assure the uniqueness of such generalized-order iterations [7, 21, 95]. For instance, Artin [7] shows that the condition of logarithmic convexity is sufficient to assure the uniqueness of the Γ-function as the extension of factorial. It may be remarked that the uniqueness of half-order iterates, or square roots, has played an essential role in Gleason's central contribution [55] toward the solution of Hilbert's fifth problem [102].

One very common and important example of the extension problem was already considered in Ex. 5.6, where conjugacy was used to define the powers x^a for arbitrary a.

As the last example, we consider the following question. Given the exponential function e^x, we have its iterates; the nth one is of course

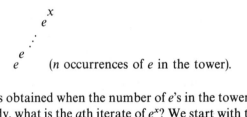

$(n$ occurrences of e in the tower).

What function is obtained when the number of e's in the tower is the complex number a? Briefly, what is the ath iterate of e^x? We start with the conjugation relation

$$e^{f(x)} = f(bx)$$

or, in simplified notation, $\exp = fbf^{-1}$. The first thing to do is to find the fixed points, that is, the roots of the equation

$$e^c = c. \tag{2}$$

There are no real roots but we have infinitely many complex ones, the nearest ones to the origin being

$$c_0 = 0.318150 + 1.337236i.$$

and its complex conjugate. This gives us the constant term in the expansion

$$f(x) = c_0 + Kx + K_2x^2 + K_3x^3, \tag{3}$$

and we also find that $b = c_0$. The next coefficient K is arbitrary, just as it was in equation (3) of Ex. 5.6; we take it to be 1. Now the other coefficients in (3) can be recursively computed; they fall off rapidly, just as in the power series for e^x. For instance, the coefficient of x^{20} is

$$-0.11163885 \cdot 10^{-11} + 0.27120056 \cdot 10^{-12}i.$$

The power series (3) may be inverted; the coefficients in f^{-1} fall off rather slowly. This is just as might be expected from the analogy with $\exp^{-1}(x) = \log x$. Once the power series expansions for $f(x)$ and for $f^{-1}(x)$ have been secured, the iteration problem is solved, and the ath iterate of e^x is $f(b^a f^{-1}(x))$. The details, such as the domains of convergence, the need for using other roots of (2) in certain vertical strips, and the optimization by choosing K properly, can be taken care of without undue difficulty.

EX. 5.8.   ERROR CONTROL IN RATIONAL COMPUTATION

What might be called the irreversible or thermodynamic aspect of conjugacy suggests that, except for some highly theoretical situations perhaps, a bypass cannot be executed perfectly, with exact return to the initial condition. Thus a bypass is usually accompanied by a loss, friction, noise, error, interest payment, etc. The present example is a method of error control or self-sharpening, which is perfect if impractical; it concerns a rare case in which the bypass-loss exists but can be completely compensated for.

Consider a nonsingular system of n linear equations

$$S: \quad \sum_{j=1}^{n} a_{ij}x_j = b_i, \quad i = 1, \ldots, n$$

with a unique solution $x_1, \ldots, x_n$. It happens often that the coefficients a_{ij} and b_i are known exactly as rational numbers; this is emphasized by writing S_{rat} instead of S. In spite of that exactness, such a system S_{rat} is solved in practice by converting first each a_{ij} and b_i to an M-place decimal so that the exact system S_{rat} is replaced by an approximating system S_{Mdec}. Then the system S_{Mdec} is solved by one of the many procedures, with M-decimal accuracy kept throughout the computations. Thus an approximate solution $x'_1, \ldots, x'_n$ is obtained, each x'_i being again an M-place decimal.

The question arises whether a solution procedure and a sufficiently large M exist, such that the exact solution $x_1, \ldots, x_n$ can be recovered from the approximate solution $x'_1, \ldots, x'_n$ alone, that is, supposing that the coefficients a_{ij} and b_i are forgotten. The answer is yes, and one might say that we use here the effective discreteness, and even finiteness, of the situation to control the error completely. Thus the basic bypass is

exact rational values   system S_{rat} $-\,-\,-\,\rightarrow$ exact solution $x_1, \ldots, x_n$

round-off $\Big\downarrow$ sharpening $\Big\uparrow$

                   solve

aproximate M-decimal   system $S_{M\text{dec}}$ $\xrightarrow{\text{solve}}$ approximate solution

values $x'_1, \ldots, x'_n.$

We prove now our assertion, choosing as the solution procedure the standard expression for x_i, which is the quotient of two determinants:

$$x_i = \begin{vmatrix} a_{11} & \ldots & a_{1i-1} & b_1 & a_{1i+1} & \ldots & a_{1n} \\ \cdot & & \cdot & \cdot & \cdot & & \cdot \\ \cdot & & \cdot & \cdot & \cdot & & \cdot \\ a_{n1} & \ldots & a_{ni-1} & b_n & a_{ni+1} & \ldots & a_{nn} \end{vmatrix} / |a_{ij}|, \ i = 1, \ldots, n. \qquad (1)$$

First, consider a determinant

$$x = A/B = |N_{ij}/D_{ij}| \qquad i,j = 1, \ldots, n$$

where each fraction is a quotient of two integers whose greatest common divisor is 1. Then there are obvious upper bounds on A and B, for instance,

$$|B| \leqslant D^{n^2}, \qquad |A| \leqslant N^{n^2} D^{n^2} \qquad (2)$$

where

$$D = \max_{i,j} |D_{ij}|, \qquad N = \max_{i,j} |N_{ij}|.$$

Next, if $x = A/B$, then the decimal development of x is a mixed periodic decimal

$$x = I.FPP \ldots \qquad (3)$$

in which I, F, and P are groups of digits, with f digits for the prefatory decimal group F and p digits in the repeating, or periodic, part P. It might happen that x is a pure repeating decimal so that F is absent (or $f = 0$), or that x is a terminating decimal so that P consists of zeros alone. Further,

$$f \leqslant |B|, \qquad p \leqslant |B|. \tag{4}$$

Upper bounds far better than (2) and (4) could be given, but we can afford to be generous. Now, the expressions (1) for the x_i are continuous in the quantities a_{ij} and b_i because the denominator determinant in (1) is by hypothesis not 0. Hence an explicit integer M can be exhibited so that with the round-offs in a_{ij} and b_i, and with all computations carried out to M digits accuracy, each approximate decimal form

$$x'_i = I_i . F_i P_i P_i \ldots$$

has the prefatory part F_i and at least two complete periods P_i exact. Since F_i and P_i can be found, it follows that the exact value x_i is recoverable from x'_i:

$$x_i = I_i . F_i + 10^{-f_i} P_i / (10^{p_i} - 1),$$

which was to be shown. Of course, we can ask now for better upper bounds, for optimal procedures, etc. In conclusion, it may be mentioned that similar error-control procedure applies to any case in which the unknowns depend rationally on parameter values that are rational.

EX. 5.9.   INTEGRAL TRANSFORMS

Conjugacy is clearly exhibited in this major subject, but since the principles are well known we limit ourselves to a few general remarks. One deals here with the function domain F and the transform domain J. The relations S and S^{-1} are given by $S: F \rightarrow J$, $S^{-1}: J \rightarrow F$. The generic form is: for $f(x) \in F$

$$\phi(y) = Sf = \int f(x) K(x,y) dx, \qquad f(x) = S^{-1}\phi = \int \phi(y) R(x,y) dy$$

with suitable kernel K and reciprocal kernel R and with appropriate paths of integration. The principal theorem for each specific integral transform asserts that, under specified conditions and in a specified sense, S and S^{-1} are indeed inverses; this is called the inversion theorem. With regard to the question of whether bypasses are lossy or loss-free, it may be remarked here that the integral-transform bypass exhibits a certain loss; it is so insignificant, for most mathematical purposes, that it is sometimes forgotten. That is: the function recovered at the last stage can differ from the original one over a set of measure zero. When it comes to applications, integral transforms offer many advantages, for instance easy handling of the initial conditions with linear differential equations. However, the principal feature of usefulness is that complicated operations in F, such as differentiation or translation or convolution, are mapped onto simple operations in J (usually onto some sort of multiplication). The bypass appears here in the following form: if $E(f) = 0$ is a complicated functional equation for f, the transform $\phi = Sf$ satisfies a simple functional equation $E_1(\phi) = 0$ and if T is the process of solving the latter, the solution of $E(f) = 0$ is obtained by applying STS^{-1} to f. Occasionally two different transforms are applied in sequence, according to the schema (C_4) in the Introduction. For instance, the Landau–Rümer solution of the cosmic-ray shower equations is achieved by the schema $S_1 S_2 T S_2^{-1} S_1^{-1}$ where S_1 in the Mellin transform and S_2 is the Laplace transform [85, 130]. In connection with the present example, the following very general question may be raised: What happens if something goes wrong with one of the three parts of the bypass? Is something still salvageable? This topic—the pathology of bypasses—will occupy us later. In the meantime we observe that even if the preceding transformed equation $E_1(\phi) = 0$ is not simple and cannot be solved, so that the middle part T of STS^{-1} fails, we may still be able to obtain some useful information about f. For, it may be possible to compute the first few derivatives $\phi^{(j)}(0)$, and with many integral transforms those values give the corresponding moments of f.

EX. 5.10. ALGEBRAIC ADDITION THEOREMS FOR TRIGONOMETRIC AND ELLIPTIC FUNCTIONS

We say that a function $\phi(x)$ has an algebraic addition theorem if there is an algebraic function $A_2(u,v)$ such that for all x and y

$$\phi(x + y) = A_2(\phi(x),\phi(y)). \tag{1}$$

This extends at once to

$$\phi(x + y + z) = A_2(\phi(x),\phi(y + z)) = A_3(\phi(x),\phi(y),\phi(z))$$

and generally to

$$\phi\left(\sum_1^n x_1\right) = A_n(\phi(x_1), \ldots, \phi(x_n)). \tag{2}$$

Here $A_3, \ldots, A_n$ are algebraic functions simply built up from A_2 of (1). A_2 is symmetric: $A_2(u,v) = A_2(v,u)$, and so is $A_n(u_1, \ldots, u_n)$ under any permutation of its n arguments. The special case known as algebraic n-tuplication theorem results if, for example,

$$x_1 = x_2 = \ldots = x_n = x;$$

(2) becomes then

$$\phi(nx) = A(\phi(x)), \qquad A(u) = A_n(u, \ldots, u). \tag{3}$$

For instance, with the specific example $\phi(x) = \sin x$, we have

$$A_2(u,v) = u\sqrt{1 - v^2} + v\sqrt{1 - u^2}$$

and generally

$$A_n(u_1, \ldots, u_n) = S_1 - S_3 + S_5 - \ldots + (-1)^{(k-1)/2}S_k,$$

where $k = 2[(n-1)/2] + 1$ is the largest odd integer $\leqslant n$, and

$$S_m = \sum u_{i_1} \ldots u_{i_m}\sqrt{1 - u_{j_1}^2} \ldots \sqrt{1 - u_{j_{n-m}}^2},$$

the summation extending over all the $\binom{n}{m}$ distinct partitions

$$\{1,2, \ldots, n\} = \{i_1, \ldots, i_m\} \cup \{j_1, \ldots, j_{n-m}\}.$$

For the n-tuplication with $\phi(x) = \sin x$, we have in (3)

$$A(u) = \sum_{j=0}^{[(n-1)/2]} (-1)^j \binom{n}{2j+1} u^{2j+1}(1 - u^2)^{(n-2j-1)/2}.$$

Equation (2) may be rewritten as

$$\phi\left(\sum_1^n \phi^{-1}(u_i)\right) = A_n(u_1, \ldots, u_n), \tag{4}$$

expressing the algebraic addition theorem in the bypass form

$$\phi\sum_1^n \phi^{-1} = A_n$$

or in words: any finite addition is conjugate under ϕ to an algebraic function. An objection of artificiality might be made here: the form (2) may appear more natural for the statement of the algebraic addition theorem than the form (4). However, it is observed that in (4) the right-hand side is free of ϕ, making it perhaps easier to deal with, at least when ϕ itself is unknown. It is a matter of historical record that the two earliest discoveries concerning elliptic functions had been made by Fagnano (in 1716–1718) and by Euler (in 1751–1753) before the elliptic functions themselves were known [95, 128]. These discoveries were precisely in the form (4), although Fagnano and Euler dealt with elliptic integrals rather than with elliptic functions. That is, instead of (4) there appears the conjugation

$$\psi^{-1}\left(\sum_1^n \psi(u_i)\right) = A_n(u_1, \ldots, u_n). \tag{5}$$

Fagnano considered the lemniscate, which is the locus of points (x,y) for which the product of distances from the points $(-a,0)$ and $(a,0)$ is the constant a^2. Following Reference 128 we take $a = 2^{-\frac{1}{2}}$; the parametric equations of the lemniscate in terms of radius vector r are then

$$x = \sqrt{(r^2 + r^4)/2}, \qquad y = \sqrt{(r^2 - r^4)/2}.$$

With the limitation to positive square roots and to points in the first quadrant, the arc length s taken from the origin $P(0)$ to the point $P(r)$ is

$$s = \psi(r) = \int_0^r \frac{dr}{\sqrt{1 - r^4}}.$$

This lemniscatic integral is a special case of an elliptic integral of the first kind, and Fagnano found an algebraic duplication theorem for it: He showed that if $\psi(r) = 2\psi(u)$, then r and u are connected by the relation

$$r^2(1 + u^4)^2 - 4u^2(1 - u^4) = 0. \tag{6}$$

In terms of (5), Fagnano's duplication is

$$\psi^{-1}(2\psi(u)) = A(u),\qquad(7)$$

where the algebraic function $r = A(u)$ is obtained from (6) as

$$A(u) = \frac{2u\sqrt{1-u^4}}{1+u^4}.\qquad(8)$$

Fagnano's work stimulated Euler to extend the duplication theorem (7) for the lemniscatic integral ψ to an addition theorem. A very plausible reconstruction of Euler's train of thought will be found in Siegel [128, pp. 7–11]. Such an addition theorem could be assumed to have the form

$$\psi(u) + \psi(v) = \psi(A(u,v));\qquad(9)$$

$A(u,v)$ is a symmetric algebraic function which reduces for $u = v$ to (8). The simplest such function is

$$A(u,v) = \frac{u\sqrt{1-v^4} + v\sqrt{1-u^4}}{1+u^2v^2},$$

and this is exactly what Euler proved for (9). Finally, Euler extended his work from the lemniscatic integral $\psi(u)$ to the general elliptic integral of the first kind. This was done in two steps, first generalizing

$$\psi(u) = \int_0^u \frac{du}{\sqrt{1-u^4}} \quad \text{to} \quad \chi(u) = \int_0^u \frac{du}{\sqrt{P(u)}}$$

where

$$P(u) = 1 + au^2 - u^4.$$

Here Euler showed that

$$\chi(u) + \chi(v) = \chi(A(u,v))$$

with

$$A(u,v) = \frac{u\sqrt{P(v)} + v\sqrt{P(u)}}{1+u^2v^2}.$$

The final extension was to the case of an arbitrary quartic polynomial $P(u)$ in the integral for $\chi(u)$. Here the computations grow somewhat involved; an indication is given in Reference 128 and a fairly complete account is in [60, pp. 23–27].

Euler's work led to the development of the subject of elliptic functions at the hands of Legendre and to the subsequent flowering of the theory of elliptic functions with Abel and Jacobi, who, unlike Legendre, realized the importance of inverting elliptic integrals into elliptic functions. A complete analogy here is replacing the integral

$$x = \int_0^y \frac{dy}{\sqrt{1 - y^2}} = \arcsin y$$

with complicated functional properties by the simple inverse function $y = \sin x$ with its characteristic periodicity property; this, of course, generalizes to the double periodicity of the elliptic functions. In connection with this inversion we quote Jacobi's motto *"Man muss immer umkehren"* (one must always invert); this has obvious importance and clear relation to bypass in general. The final theorem tying together algebraic addition theorems with the elliptic functions is due to Weierstrass: A function $f(z)$ is elliptic if and only if it has an algebraic addition theorem (here exponential and trigonometric functions count as special degenerate cases of elliptic functions).

Chapter Six

Examples in Algebra

EX. 6.1. SIMILARITY OF MATRICES

Let V be an n-dimensional vector space over some field F and let W and T be two $n \times n$ matrices with entries in F. Two apparently unrelated questions are asked, one theoretical and the other practical: When do W and T represent the same linear transformation of V? How can we raise W to some moderately large positive integer power m conveniently, that is, with possibly few multiplications?

To answer these questions one uses the elements of vector-space theory, in particular, the representation of a linear transformation by reference to a basis, and the change of bases. In passage from one basis to another the vector X of coordinates in the old basis is transformed to the vector X_1 of coordinates in the new basis by $X_1 = XS$ where S is a nonsingular, and hence invertible, $n \times n$ matrix. Any linear transformation of V, from old coordinates X to some new coordinates Y, is described by $Y = XA$, where A is an $n \times n$ matrix. We indicate these facts by the diagrams [16]:

$$X \xrightarrow{\ S\ } X_1, \qquad X \xrightarrow{\ A\ } Y.$$

Now, our first question becomes, in slightly greater detail: When do W and T represent the same linear transformation of V, written down relative to two bases of V?

Suppose that with respect to one basis of V we have $Y = XW$ and with respect to another $Y_1 = X_1 T$. Then by the preceding we get the bypass diagram

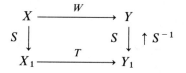

so that

$$XW = Y = Y_1 S^{-1} = X_1 T S^{-1} = XSTS^{-1}$$

and hence indeed $W = STS^{-1}$.

This also answers the second question: Given W, produce S so that T is possibly simple, in the sense of behaving as closely as possible like a scalar. Here the schema (C_3) from Chapter 1 appears, and we have then

$$W^m = ST^m S^{-1}.$$

Now S is found so as to make T most scalarlike or diagonal, if this is possible. Otherwise, assuming suitable factorization of polynomials over F, S is found which puts T in the Jordan canonical form [16]. The matters are simplest when F is the field C of complex numbers. The principal theorem is then: A necessary and sufficient condition for W to be similar to a diagonal matrix T is that the eigenvalues $\lambda_1, \ldots, \lambda_n$ of W be all distinct, and then

$$T = \begin{bmatrix} \lambda_1 & 0 & \ldots & 0 \\ 0 & \lambda_2 & \ldots & 0 \\ \cdot & \cdot & & \cdot \\ \cdot & \cdot & & \cdot \\ \cdot & \cdot & & \cdot \\ 0 & 0 & & \lambda_n \end{bmatrix}.$$

Otherwise, if $\lambda_1, \ldots, \lambda_q$ are the distinct eigenvalues of W, while λ_{q+i} has muliplicity n_i, $i = 1, \ldots, s$, W is similar to the Jordan matrix

$$
T = \begin{bmatrix}
T_0 & 0 & . & . & . & . & 0 \\
0 & T_1 & & . & . & . & 0 \\
. & . & & & & & . \\
. & . & & & & & . \\
. & . & & & & & . \\
0 & 0 & . & . & . & . & T_s
\end{bmatrix}
$$

where T_0 is a $q \times q$ diagonal matrix

$$
\begin{bmatrix}
\lambda_1 & 0 & . & . & . & . & 0 \\
0 & \lambda_2 & & . & . & . & 0 \\
. & . & & & & & . \\
. & . & & & & & . \\
. & . & & & & & . \\
0 & 0 & . & . & . & . & \lambda_q
\end{bmatrix}
$$

while T_i is an $n_i \times n_i$ matrix of the form

$$
\begin{bmatrix}
\lambda_{q+i} & 1 & 0 & & . & . & . & . & 0 \\
0 & \lambda_{q+i} & 1 & & . & . & . & . & 0 \\
. & . & \lambda_{q+i} & & & & & & . \\
. & . & . & & & & & & . \\
. & . & . & & & & & & 1 \\
0 & 0 & 0 & & & & & & \lambda_{q+i}
\end{bmatrix}
$$

It may be observed that the bypass-stacking schema (C_4) from Chapter 1 also appears in the process: The reduction of W to the diagonal or Jordan form T need not be achieved all at once but may be accomplished sequentially by the so-called block-reduction process. That is, a nonsingular S_1 is found first, such that

$$
W = S_1 T_1 S_1^{-1} = \begin{vmatrix} U_1 & 0 \\ 0 & U_2 \end{vmatrix}
$$

and then the reduction process is continued on the square matrix blocks U_1 and U_2. Eventually one obtains

$$
W = S_K \ldots S_1 T S_1^{-1} \ldots S_K^{-1} = STS^{-1}
$$

where $S = S_K \dots S_1$ and T is of the desired form. Another useful observation hinges on the schema (C_3) again. Let

$$f(x) = a_0 + a_1x + a_2x^2 + \dots = \sum_0^\infty a_k x^k$$

be a suitably convergent power series, and suppose that the matrix function $f(W)$ is defined by its Neumann series [33]

$$f(W) = \sum_0^\infty a_k W^k,$$

where $W^0 = I$ is the $n \times n$ unit matrix. Then $W = STS^{-1}$ and $W^k = ST^kS^{-1}$ imply that

$$f(W) = f(STS^{-1}) = Sf(T)S^{-1}$$

Of special interest are the cases $f(x) = \exp x$ and $f(x) = \log(1 + x)$. These and similar examples have applications to systems of differential equations [31] and to the Fredholm and Volterra integral equations [33].

EX. 6.2.   CONJUGACY IN GROUPS

In group theory, conjugacy appears in the following basic concepts: conjugate elements, conjugate subgroups, inner automorphisms. Let G be a group; $a, x, q, \dots$ its elements; $H, U, V, \dots$ its subgroups. Then x and y are conjugate if $x = aya^{-1}$ for some a in G; U and V are conjugate if $U = aVa^{-1}$ for some a in G; and an inner automorphism is a mapping of G into itself given by $x \rightarrow axa^{-1}$. These three notions are quite fundamental in group theory. For instance, the central concept of a normal subgroup can be defined by conjugacy. First, H is a normal subgroup of G if and only if it coincides with every subgroup of G which is conjugate to H [84]. Then, a subgroup H of G is normal if and only if, together with each of its elements h, it also contains all elements of G, conjugate to h [84]. Note further that x and y in G commute if and only if either is self-conjugate under the other. So, the notions of normalizers, centralizers, and centers also can be expressed by conjugacy.

Beside such structural uses, conjugacy also enters into combinatorial, or counting, arguments [68]. For instance, since element conjugacy is an equivalence relation, it partitions G into disjoint equivalence classes. For a finite group G, a valuable identity results by counting the elements of G in two different ways. First, this is done directly, resulting in ord G (the order of G, i.e., the number of its elements). Then, one counts the number of elements in a conjugacy class and sums over the classes. Equating the two counts results in the class equation of the group [68], [142]:

$$\text{ord } G = \sum \text{ord } G/\text{ord } N(a).$$

Here $N(a)$ is the normalizer of a in G, that is, the collection of all x in G such that a is self-conjugate under $x : a = xax^{-1}$. The summation runs over any set of elements with exactly one element in each class. From the class equation one gets simple proofs of a number of important theorems, for instance, of Sylow's theorem that if p is a prime and $p^k|\text{ord } G$, then G has a subgroup of order p^k.

Probably the most important single idea in group theory is that of a normal subgroup, also known as a normal divisor or a self-conjugate subgroup. This also marks the earliest algebraic appearance, though implicitly, of conjugacy. In some ways a normal subgroup of a group corresponds to a divisor of an integer, just as group theory generally, at least for finite or countable groups, corresponds to the elementary number theory. However, the latter works with the very special system, the integers, having a rich structure, whereas the former applies to a very wide class of systems, works with far fewer tools, and therefore is much harder as well as more general.

Normal subgroups were introduced by E. Galois who actually used the coset partitions of G with respect to a subgroup H and singled out as proper (nowadays called regular) those partitions for which the right and the left cosets coincide. But it is well known [84] that a necessary and sufficient condition for that is that H be normal (i.e., self-conjugate) in G.

Since so much in group theory depends on conjugation it may be asked whether everything does. That is, can the whole of group theory be re-axiomatized by bypass? Conversely, what algebraic group-like structures arise when some sort of "bypass" system is introduced with the axiomatization of some of the properties, such as (C_3), (C_4), and (C_5), from Chapter 1? Let us examine first what is needed to express (C_3), (C_4), and (C_5) rigorously. Since products like W_1W_2, T_1T_2, S_1S_2 occur, some "ordinary" multiplication is needed first. So, we introduce it, but without any additional requirements such as solving equations of the type $ax = b$, or the existence of inverses or units, or asking for associativity. Along with the "ordinary" multiplication, the bypass operation itself is introduced as the second type of multiplication. That is, we work with a (nonempty) system B of elements, closed under two binary operations ab and (a,b): For every ordered pair of elements a, b in B, there is a unique product ab and a unique bypass (a,b) in B.

The intended interpretation of (a,b) is, of course, aba^{-1}; the question is how to axiomatize toward this interpretation. This axiomatization will be done, so to speak, in the open. That is, the axioms can be illustrated and motivated even in the crude graphical language of digging shafts and tunnels, as in Chapter 1. Of course, this illustrative power is purchased at the price of having two multiplications rather than one. Thus, from a formal point of view, there is a redundancy, and perhaps even an unnecessary redundancy. Still, bypasses stand to benefit from standard group theory, even if the reverse is not necessarily true.

First (C_3), (C_4), and (C_5) lead to:

(E)   for every a,b and x $(a,x)(a,y) = (a,xy)$,
(D)   for every a,b and x $(a,(b,x)) = (ab,x)$,
(S)   for every b there is an x such that for every a, $a = (b,(x,a))$.

No comment is necessary regarding (E), which corresponds to extending of bypasses in length, or regarding (D), which corresponds to extending them in depth or stacking them; (E) and (D) mirror (C_3) and (C_4) exactly.

The weak symmetry, or rather, the weak inversion axiom (S) arises from (C_5). Start with the bypass relation $a = (b,c)$ stating that a is conjugate to c under b. We should like to express (C_5) by saying that therefore c is conjugate to a under b^{-1} since for a group $a = bcb^{-1}$ implies $c = b^{-1}ab$. As there are no inverses (yet), we settle for a weaker form in which c is conjugate to a under some x that depends on b only. Since now

$$a = (b,c) \qquad \text{and} \quad c = (x,a),$$

(S) is obtained by substituting for c from the second equation into the first one. The dependence of x on b alone follows from the order of the quantifiers in (S).

How much of group structure can be extracted from the axioms (E), (D), (S)? Clearly not all of it, for if the group happens to be commutative then our intended interpretation

$$(a,b) = aba^{-1} = b$$

collapses every bypass to its second term. But then (E), (D), and (S) become vacuous. Hence some structure is to be expected from these three axioms only if our system B has a degree of noncommutativity, thus allowing the axioms to take hold on every element in B. A fairly simple and general way to force this condition in a group G would be to require that no element, other than the identity 1, commutes with every x in G. That is, the center $Z(G)$ of G is trivial: $Z(G) = \{1\}$. Luckily, it turns out that this condition is simply expressible in bypass terms as a weak right cancellation law:

(R)   if for all x $(a,x) = (b,x)$,   then $a = b$.

For, in a group G and with $(a,x) = axa^{-1}$, we get from (R)

$$axa^{-1} = bxb^{-1} \text{ for all } x$$

so that

$$x(a^{-1}b) = (a^{-1}b)x \text{ for all } x,$$

and hence $a^{-1}b \in Z(G)$, but $Z(G) = \{1\}$ and so $a = b$. It turns out, interestingly, that the axiom (E) is not needed. Accordingly, a bypass system is defined formally to be a nonempty set B, with two binary operations ab and (a,b), satisfying the rules (D), (S), and (R). It will be shown next that a bypass system B is necessarily a group under its multiplication ab.

First, the multiplication is shown to be associative. Take a, b, c, x to be arbitrary and consider

$$(\overline{abc}, x) \quad \text{and} \quad (\overline{abc}, x),$$

the bar denoting merely the grouping of elements. By repeated use of (D),

$$(\overline{abc}, x) = (ab,(c,x)) = (a,(b,(c,x))),$$
$$(\overline{abc}, x) = (a,(bc,x)) = (a,(b,(c,x))),$$

so that

$$(\overline{abc}, x) = (\overline{abc}, x) \quad \text{for all } x$$

and therefore the associative law

$$\overline{abc} = \overline{abc}$$

follows on applying (R). We remark that the level stratification of the bypass enters here very essentially; in particular, it makes the preceding demonstration of associativity so particularly easy. The existence of a unique right identity e is shown next. Let b be an arbitrary element; by (S) there is some x, depending on b only, such that

$$a = (b,(x,a)) \quad \text{for all} \quad a.$$

An application of (D) puts it in the form

$$a = (bx,a) \quad \text{for all } a.$$

This x is, of course, the intended inverse b^{-1} of b; x itself is not necessarily unique but the product bx is, because

$$(bx_1,a) = (bx_2,a) \quad \text{for all } a$$

implies $bx_1 = bx_2$ by (R). Also, if another element c is substituted for b, and the hypothetical inverse x of b goes over into the hypothetical inverse u of c, then

$$(bx,a) = (cu,a) \quad \text{for all} \quad a$$

so that $bx = cu$ by (R). This shows the existence of a unique distinguished element e in B, such that

$$a = (e,a) \quad \text{for all } a.$$

To show that e is the desired right identity, consider the element

$$(a,(e,b)), \quad a,b \quad \text{arbitrary,}$$

and reduce it in two different ways. Since $(e,b) = b$, the element is (a,b), but by (D) it is also (ae,b), so that

$$(a,b) = (ae,b) \qquad \text{for all} \quad b$$

and therefore $a = ae$ by (R), which was to be shown. Once e has been shown to be the right identity, the hypothetical inverse x of b becomes an actual right inverse since now $bx = e$.

However, it is well known [61, 84] that an associative multiplication together with right identity and right inverse, which need not even be unique, imply the full group axions. This completes the proof. It might be of interest to know what groups can occur as multiplicative groups of bypass systems. Obviously, any group with trivial center is admissible: We define $(a,b) = aba^{-1}$ and verify the axioms. Such groups are known to include symmetric groups S_n for $n > 2$ and all finite noncommutative simple groups.

EX. 6.3.   MISCELLANEOUS ALGEBRAIC OCCURRENCES

Conjugacy also occurs in algebra in systems other than matrices or groups. For instance, it plays an important role in some proofs of Wedderburn's theorem, in the proof of Jacobson's theorem, and in the very statement of the Brauer–Cartan–Hua theorem [68]. The first of these asserts that a finite division ring is a field. The second one states that if R is a ring and every element a of R satisfies an equation

$$a^n = a, \qquad n = n(a) > 1,$$

then R is a field. The third one asserts that if D is a division ring and K a division subring of D such that

$$xKx^{-1} \subset K \text{ for every } x \neq 0 \text{ in } D,$$

then either $K = D$ or $K \subset Z(D)$.

Finally, conjugacy appears to turn up in a branch of algebra called the category theory, a subject found to be rather esoteric by many working mathematicians. It has some similarity to the last two novels of James Joyce. Certainly both Joyce and it are hyperconcerned with language; according to enthusiasts both achieve remarkable compression of meaning, while according to opponents both bypass it. So far as can be judged by outsiders, the general opinion is that category theory has one rather interesting concept, the so-called adjoint functor. This view seems to be supported by the practitioners; for example, in the preface to Reference 92, we read: "Next comes the fundamental idea of an adjoint pair of functors. ... The slogan is 'Adjoint functors arise everywhere.'" Indeed they do, being bypasses, though dressed up and served forth exotically; for instance, in the quoted source [92, p. 79] there is the following definition of the "right adjoint" G:

$$\phi^{-1}(GK \circ g) = k \circ \phi^{-1}g$$

where the symbols stand for objects, natural bijections, etc.; applying ϕ, we would then have

$$GK \circ g = \phi k \circ \phi^{-1} g.$$

EX. 6.4.   STRUCTURE BYPASS

Here we wish to summarize a general process that appears to occur in many branches of mathematics and at many levels of mathematical sophistication. A convenient starting point is the Euler principle of structural parameter discussed in Chapter 2. This depends on observing first that in a mathematical situation, *as it is given initially*, there may be insufficient structure for proper resolution of the difficulty faced. More simply put, there are not enough variables. So, an additional structural (or, we might say, structurating) parameter is suitably and shrewdly introduced, one then works in the extended context, and at the end the additional parameter is removed (or "killed").

For the present introductory purpose the main point is not the parameter itself, nor even the technique of handling it, but the benefit conferred by it of bringing in additional structure. The thing is that such benefit may arise in other, much more sophisticated, ways. Thus the present example might be roughly and preliminarily said to concern something like a generalized Euler principle. This might be phrased: How to produce by mathematical ingenuity and novelty, an extended structure which is sufficiently rich for proper resolution of a difficult mathematical situation.

Three objections arise at once. First, the preceding formulation is far too wide and general; as it stands, it includes an uncomfortably large fraction of the whole of mathematics and therefore is useless. Second, why include the subject with algebra? Third, where does bypass come in? The first objection is granted unconditionally; the problem is then how to obtain a useful restriction. A hint which points toward such restricting is obtained by considering the second objection. It will turn out in a number of important instances of our example that the extended structure depends on introducing a new and deep sort of "multiplication" or "division", or perhaps both. Thus it should not be surprising that all our instances concern groups in one way or another. And in view of Ex. 6.2, as well as on other grounds, it is almost natural that somehow or other bypass should enter into the proceedings (e.g., introduce the new structure—work in the extended context—return to the original situation).

After the preceding generalities and abstractions our first instance may appear to be distressingly simple and concrete: Ex. 4.0, concerning the work of Thales on proportion and similarity. Here the new multiplication and division are *the ordinary ones*. The object of Thales was, as we recall, to find the height x of the inaccessible pyramid, given the length c of its accessible shadow, and the height a and shadow length b of an accessible object (e.g., a vertical rod or tree). Thalesian proportion and similarity produced sufficient structure to solve the problem geometrically, as is shown in Figs. 1a–1c of Ex. 4.0. We say that Thales was able to solve the problem *geometrically* because, for one thing, he could not have solved it otherwise in his day and age. Yet his geometry, as we shall argue, contains the seed of the algebra that is to come.

To us the problem of Thales is trivially settled and solved by the mere *writing* of the equation $x = ac/b$. But how would one solve it without a clear concept of multiplication, division, or equations? And yet, it may be claimed with some justice that Thales achieved those concepts, and even that he discovered, however implicitly, the multiplicative group structure of (positive) real numbers. The bypass nature of the work of Thales was claimed, on geometrical grounds, already in Ex. 4.0. This might be accentuated by rewriting the equation $x = ac/b$ fancifully as $x = M_a c M_b^{-1}$. In words, and with what we have called in Chapter 3 the performative interpretation of bypass, this is a recipe for obtaining the unknown height x: To get x multiply by a the number c and then divide by b.

The preceding first instance concerned solving a very simple equation, essentially $ax = b$, but under very stringent and primitive conditions; the "extended" structure amounts to ordinary multiplication and division. The next instance is Galois theory, which concerns solving polynomial equations in general. Here the "extended" structure will amount to multiplication and division in their most general setting, that is, in the setting of a group. A phrase of C. Jordan may be quoted here: "Group theory is the metaphysics of the theory of equations." In a similar spirit, it might be said that multiplication and division are the epistemology of the Thalesian theory of proportion.

The interest now is in solving equations of the form $p(x) = 0$, where $p(x)$ is a polynomial with coefficients in some field F. Originally, of course, F was the field of rational, or real, or complex numbers. As is well known, for equations of degree $\leqslant 4$ algebraic processes have long been known to give roots in terms of the coefficients. That is to say, the roots of such equations are obtainable from the coefficients by the four arithmetic operations together with the taking of radicals (i.e., root extraction). It is noted, incidentally, that the algebraic solving process for quadratics, cubics, and quartics may be put in the form of a bypass recipe. For instance, for the quadratics we have the stacked bypass $S_1 S_2 R S_2^{-1} S_1^{-1}$ given by general quadratic $\qquad$ two roots $x_1, x_2 = \dfrac{-b \pm \sqrt{b^2 - 4ac}}{2a}$

$ax^2 + bx + c = 0$

$\downarrow$ $S_1 : ax = u$ $\qquad\qquad$ $\uparrow$ S_1^{-1}

reduced quadratic $\qquad$ two roots $u_1, u_2 = -b/2 \pm \sqrt{b^2/4 - ac}$

$u^2 + bu + ac = 0$

$\downarrow$ $S_2 : u + b/2 = v$ $\qquad\qquad$ $\uparrow$ S_2^{-1}

pure quadratic $\qquad$ solve by $\qquad$ two roots

$v^2 = b^2/4 - ac$ $\xrightarrow{\text{root extraction } R}$ $v_1, v_2 = \pm \sqrt{b^2/4 - ac}.$

Similar, though more complicated, bypasses can be given for cubics and quartics.

On the other hand, the solution of quintics or higher-degree equations was a very refractory problem, in spite of several centuries of work by many mathematicians. Such work was done on the positive hypothesis: It was believed that sufficiently complex algebraic processes would solve quintics too. It was only around the year 1800 that the contrary idea seems to have occurred to mathematicians: that such algebraic solution processes may be impossible.

Of course, showing that something cannot be done is usually harder than showing that it can be done: The impossibility proofs often call for radically new concepts and ideas. Following the earlier and apparently incorrect work of Paolo Ruffini in 1799 [27, Vol. 4] on the impossibility of solving the general quintic algebraically, such impossibility was proved by N. H. Abel in 1824. The work was published at Abel's own expense and reprinted in 1826 by Crelle in the first volume of Crelle's *Journal.* However, Abel's work was also incorrect, and it was only completed and corrected after Abel's death in 1826.

It is of interest to summarize briefly the spirit of Abel's method; this concerns the previously mentioned "positive hypothesis" that sufficiently complex algebraic processes will solve the quintic. Abel used such a complexity idea, but in reverse, so to speak; he started by assuming that a root of the general quintic *was* algebraically expressible in terms of the coefficients. Then he considered that algebraic expression which is algebraically least complex, and finally he proved a contradiction: The hypothetically least complex expression could be simplified even further. Thus Abel's work is in the spirit of what we have called in Chapter 2 the principle of stratification into layers by increasing complexity. It is claimed by Ritt in [121] that the work of Abel has in fact initiated that principle, and that it inspired Liouville to produce his complexity theory of integration in finite terms. It is even just barely possible that Abel's method may have suggested to Liouville ideas that led to the discovery by Liouville of the existence of transcendental numbers, in 1844. The main importance of the works of Ruffini and Abel is nowadays granted to be this: that they inspired Galois, who has definitively resolved the whole problem of algebraic solvability of polynomial equations.

This he accomplished by working in the context of a group together with its normal (or self-conjugate) subgroups. It is just this that brings in the additional structure of multiplication and division: The group corresponds to an integer, a normal subgroup to a divisor of that integer, the factor group to the quotient of the integer by the divisor, the chain of groups occurring in the definition of a solvable group to something like the Euclidean algorithm for divisibility, a simple group to a prime number, etc. In particular, by means of automorphisms that leave certain elements fixed, the group-with-normal-subgroups structure corresponds to a field-with-normal-extensions structure. It is precisely that correspondence which allows us structural-bypass round-trips between the level of fields and the level of groups. Fields, of course, are the proper setting for making precise the idea of algebraic solvability of $p(x) = 0$ by radicals: Starting with the field $F_0 = F$ of coefficients, produce a sequence of successive field extensions $F_0, F_1, \ldots, F_N$ by the adjunction of surds, so that F_N is the splitting field for $p(x)$. Groups, on the other hand, on account of their multiplication and division properties, are the proper setting in which to examine the structure of such extensions by surds: When the field F_{i+1} arises from F_i by adjunction of $a^{1/S}$, then the corresponding group G_{i+1} is a normal subgroup of G_i, with an Abelian factor group. Eventually, one obtains the principal theorem of Galois theory: $p(x) = 0$ is solvable by radicals if and only if its Galois group G is solvable, that is, if a chain $G_0, G_1, \ldots, G_N$ of groups exists, such that G_{i+1} is a normal subgroup of G_i with an Abelian factor group, and G_0 is G while G_N consists of identity alone.

One important observation must be added. Nowadays we define the Galois group G of an equation $p(x) = 0$ as the group of those automorphisms of the splitting field that preserve the ground field elementwise. But Galois defined G otherwise, in terms of permutations. Let the roots of $p(x) = 0$ be distinct, then G consists of all permutations of those roots, under which a correct polynomial equation (over the ground field) involving those roots goes over into another such equation. Equivalently, G consists of all those permutations of the roots which map each root onto a conjugate of it. Thus the chains $G = G_0, G_1, \ldots, G_N$ or $F = F_0, F_1, \ldots, F_N$ might be said to accomplish successive symmetry reductions. This reminds us that groups are not only to multiply and divide in, but also to measure the degree of algebraic or geometrical symmetry with.

To see how Galois theory behaves "in action," that is, how it enables us to resolve difficulties by means of structure bypasses between groups and fields, let us consider the following problem. Let $p_1, \ldots, p_n$ be n distinct primes, to show that the field $Q_1 = Q(\sqrt{p_1}, \ldots, \sqrt{p_n})$ is of exact degree 2^n over the field Q of rational numbers (it is trivially of degree $\leqslant 2^n$). The object is to show that successive adjunctions of square roots behave in a suitable sense like multiplications by 2. Q_1 is a normal, or Galois, extension of Q since it contains $\sqrt{p_i}$ and $-\sqrt{p_i}$ for every i and so it is the splitting field for the polynomial $(x^2 - p_1) \ldots (x^2 - p_n)$. An element of the Galois group G of Q_1 maps any $\sqrt{p_i}$ onto itself or onto its conjugate $-\sqrt{p_i}$. Therefore, G is a subgroup of the multiplicative group of $n \times n$ diagonal matrices with 1 or -1 as diagonal entries; hence it is a subgroup of the direct product of at most n cyclic groups of order 2. Thus G is a direct product of m such cyclic groups, with $m \leqslant n$. So it has as many subgroups of index 2 as of order 2, $2^m - 1$ of them. Since the subgroups of index 2 are in direct 1:1 correspondence with quadratic subfields, it is only necessary to show that there is the full quota of such quadratic subfields in Q_1, *that is,* $2^n - 1$. Let S_1 and S_2 be two nonempty, disjoint, complementary subsets of the set of all our n primes, and let P_1 and P_2 be the products of the primes in S_1 and in S_2; we observe that there are precisely $2^n - 1$ such decompositions. But the quadratic subfields $Q(\sqrt{P_1})$ and $Q(\sqrt{P_2})$ are distinct, since otherwise $\sqrt{P_1}$ and $\sqrt{P_2}$ would have been rationally related, which is impossible by the unique factorization of integers.

We finish this instance with the following note. What has been called structure bypass, depends on starting with an initial situation concerning one type of structure, passing over to a situation with an additional different structure, working within the extended context to resolve the difficulty, and then returning. It may happen that the complete round trip does not go through; for instance, the difficulty cannot (so far) be resolved even in the new context. But even then, the very fact of translation of the problem, from one type of structural framework into another different one, may be of great mathematical interest. A remarkable example of such a translation concerns one of the greatest unsolved questions in topology, the Poincaré conjecture. This states that a compact, connected, simply connected 3-manifold is a topological 3-sphere. A *purely group-theoretic* restatement of it has been found by Stallings and Jaco [91, p. 195].

The third instance concerns the fifth one on Hilbert's famous list of 23 unsolved problems, presented at the International Mathematical Congress in 1900: "Lie's concept of continuous transformation-groups without the assumption of the differentiability of the functions defining the group" [101, p. 68]. In modern terminology the problem, or rather a part of it, is as follows. Let G be a topological group, that is, both a group and a Hausdorff topological space, so that the group operations of product and of inverse are continuous in the topology. Suppose further that G is locally Euclidean: A neighborhood of the identity e of G is homeomorphic under a mapping ϕ to the n-dimensional Euclidean space E^n. Of course, the conjugacy xUx^{-1} shows that a neighborhood of every element of G is a homeomorph of E^n. As in Ex. 4.10, the property of being locally Euclidean allows a coordinatization of G: If x and y are in G, z is the group product xy, and x^{-1} the inverse of x, then

$$x = x(x_1, \ldots, x_n), \qquad y = y(y_1, \ldots, y_n), \qquad z = z(z_1, \ldots, z_n),$$
$$x^{-1} = x^{-1}(u_1, \ldots, u_n),$$

where

$$z_i = f_i(x_1, \ldots, x_n, y_1, \ldots, y_n), \qquad u_i = g_i(x_1, \ldots, x_n), \qquad i = 1, \ldots, n.$$

The functions f_i and g_i are continuous by hypothesis; the question is whether a homeomorphism ϕ exists that makes them (real) analytic. That is, is every locally Euclidean topological group a Lie group?

The problem, then, is to show that in a topological group the algebraic structure interacts so strongly with the topological structure that the minimal smoothness assumption of continuity forces the maximal smoothness conclusion of analyticity. An analogous, though very much more elementary, problem occurs for ordinary analytic functions $f(z)$. Here the topological structure of the plane, in particular the existence of the limit of $[f(z) - f(z_0)]/(z - z_0)$ under *all* modes of approach of z to z_0, reinforces the assumption of differentiability to the conclusion of analyticity.

The crux is again that there is not enough structure initially, the additional structure is again a new group structure, and this additional structure is introduced as follows. The conjugacy xUx^{-1} enables one to work near the identity e of G. It is shown that a neighborhood of e contains no subgroups (except for the trivial one, $\{e\}$). Then it is shown that a neighborhood of e exists, every element of which has a unique square root. Since there are now unique square roots, there also exist unique roots of every order 2^n, $n = 0, 1, \ldots$. Hence, by using dyadic fractions and by a passage to the limit, it is shown that every element x in the neighborhood lies on a one-parameter group germ $x(t)$, $t \geqslant 0$, issuing from the identity $e = x(0)$. Once a neighborhood of e is filled by such germs, a new operation $x \circ y$ is defined by

$$\lim_{m \to \infty} [x(t/m)y(t/m)]^m$$

in terms of the old group-operation xy. The new operation behaves like an addition; together with xy, it enables one to attach a certain n-dimensional vector space to G at e, and by the conjugacy xUx^{-1} also at every point of G. Finally, a basis of that space leads to an analytic coordinatization of G.

The last instance concerns the work of S. Novikov on generalized torsion of topological manifolds, for which he was awarded the mathematical equivalent of the Nobel Prize [74, Vol. 1, p. 11]. This is a deep problem concerning a far-reaching generalization of the Gauss–Bonnet theorem of classical differential geometry. The problem is about certain cohomological constructs, called Pontryagin classes, which generalize the concept of total curvature from ordinary closed surfaces to certain compact simply connected manifolds M. The object is to show that these constructs are topological invariants. The structure bypass is very clear here: Novikov starts with M, then he complicates the algebraic structure of M by "un-simply-connecting" it, he works then with a related manifold M_1, which, unlike M, has a nontrivial fundamental group, and eventually he obtains the desired result by returning to M.

Chapter Seven

Miscellaneous Mathematical Examples

EX. 7.1. GENERATING FUNCTIONS AND SELF-RELATING

A simple typical problem here is to find the Fibonacci numbers $f_0, f_1, f_2, \ldots$. These are given by the initial values

$$f_0 = 1, \qquad f_1 = 1 \tag{1}$$

and thereafter by the recursion

$$f_{n+2} = f_{n+1} + f_n, \qquad n = 0, 1, 2, \ldots. \tag{2}$$

It follows from (1) and (2) that each f_n is a definite positive integer and the problem is to represent the general term f_n as an explicit function of n. Thus there is a considerable formal resemblance to the problem of explicit iteration from Ex. 5.7. The method of generating functions starts with introducing the power series

$$f(x) = f_0 + f_1 x + f_2 x^2 + \ldots. \tag{3}$$

Here (1) implies that

$$f_2 x^2 + f_3 x^3 + \ldots = f(x) - x - 1,$$
$$f_1 x^2 + f_2 x^3 + \ldots = x f(x) - x,$$
$$f_0 x^2 + f_1 x^3 + \ldots = x^2 f(x).$$

Now (2) shows that the first series is the sum of the other two:

$$f(x) - x - 1 = xf(x) - x + x^2f(x) \tag{4}$$

so that

$$f(x) = \frac{1}{1 - x - x^2}. \tag{5}$$

To finish the problem, we decompose the generating function (5) into partial fractions

$$f(x) = \frac{A}{1 - ax} + \frac{B}{1 - bx}, \tag{6}$$

where

$$a = (1 + \sqrt{5})/2, \quad b = (1 - \sqrt{5})/2, \quad A = (5 + \sqrt{5})/10, \tag{7}$$
$$B = (5 - \sqrt{5})/10$$

and so f_n is explicitly given as

$$f_n = \frac{1}{\sqrt{5}}\left[\left(\frac{1 + \sqrt{5}}{2}\right)^{n+1} - \left(\frac{1 - \sqrt{5}}{2}\right)^{n+1}\right]. \tag{8}$$

Is there a bypass here and if so, where? As in some other examples, it is hidden by its very obviousness and is easily missed altogether. The first part U of this bypass UTU^{-1} stands for the extremely important operation of uniting an infinitude of objects at one level so as to form a single object at the other level. In more detail, the infinitely many separate members of the sequence $f_0, f_1, f_2, \ldots$ are transformed by U into the one generating function $f(x)$ in (3). The importance of U lies in achieving a *closure* that shows itself in the next step T of the bypass. At the individual level the given recursion (2) only relates *certain objects to some other objects*. But at the collective level the closure operates so that (2) reflects into (4): $f(x)$ *is related to itself.* That is, a functional equation (4) is obtained for the unknown generating function $f(x)$. Finding this equation (4) and solving it, as in (5), forms the middle part T of our bypass; the ease of performing T is due precisely to the closure performed by U. The last part U^{-1} is the reverse of U: While U denotes the *uniting* of the sequence $f_0, f_1, f_2, \ldots$ into the single object $f(x), U^{-1}$ denotes the *separating* of $f(x)$ back into the infinitude $f_0, f_1, f_2, \ldots.$ This separating is easily accomplished by producing the partial fractions decomposition (6), which results in (8).

The Fibonacci numbers could also be found by a different, though similar, bypass $U_1 T_1 U_1^{-1}$. This one is based on a different way U_1 of putting the sequence $f_0, f_1, f_2, \ldots$ together to form a function. Instead of the ordinary generating function (3), we use the exponential generating function

$$g(x) = f_0 + f_1 x/1! + f_2 x^2/2! + \ldots. \tag{9}$$

It follows now by differentiation that

$$g'(x) = f_1 + f_2 x/1! + f_3 x^2/2! + \ldots,$$
$$g''(x) = f_2 + f_3 x/1! + f_4 x^2/2! + \ldots,$$

and on comparing the last three series, the recursion (2) shows that here

$$g''(x) = g'(x) + g(x). \tag{10}$$

T_1 stands for the process of obtaining the preceding relation between g and its derivatives and for solving the resulting differential equation (10). The solution is of the form

$$g(x) = C_1 e^{m_1 x} + C_2 e^{m_2 x},$$

where m_1 and m_2 are the roots of the quadratic

$$m^2 = m + 1,$$

and the constants C_1, C_2 are determined by the initial conditions

$$g(0) = 1, \qquad g'(0) = 1$$

obtained from (1) and (9). It is found that

$$m_1 = a, \qquad m_2 = b, \qquad C_1 = A, \qquad C_2 = B,$$

where a, b, A, B are given by (7), so that

$$g(x) = A e^{ax} + B e^{bx}. \tag{11}$$

Although the middle part T_1 is perhaps somewhat harder than the previous middle part T, the last part U_1^{-1} is now easier than U^{-1}. Because of the simplicity of the exponential series, (11) shows at once, without anything like the partial fractions used before, that

$$f_n = Aa^n + Bb^n,$$

which is the same as (8) before. So, it might be said that the two methods of solving our problem, UTU^{-1} and $U_1 T_1 U_1^{-1}$, provide an example of trading off difficulty between the middle and the outer parts of bypasses. More essentially, this example emphasizes in a modest way a problem that arises with bypasses in general: If several bypasses work, how to choose the best one? A tentative answer is, perhaps, to apply them all and to compare their functioning, somewhat in the spirit of the two-ways principle of Chapter 2.

In the preceding example involving Fibonacci numbers the ordinary and the exponential generating functions work about equally well. However, it is easy to give examples where this is not so: Only one or the other works or, at any rate, works conveniently. The situation is quite similar to the Ex. 5.9 on integral transforms where it was stated that difficult operations at one level map onto easy operations at the other level. Here, too, a special role is played by the structure of multiplications at the two levels. Thus there may be a certain relation to the structure bypass from Ex. 6.4.

Suppose, for instance, that there are two sequences

$$\mathscr{A} = (a_0, a_1, a_2, \ldots), \qquad \mathscr{B} = (b_0, b_1, b_2, \ldots) \tag{12}$$

and a recursion involving the general expression

$$a_0 b_n + a_1 b_{n-1} + \ldots + a_{n-1} b_1 + a_n b_0, \qquad n = 0, 1, 2, \ldots. \tag{13}$$

There is then the Cauchy convolution sequence: a product $\mathscr{A} * \mathscr{B}$, which is the sequence $(a_0 b_0, a_0 b_1 + a_1 b_0, \ldots)$ whose nth term is given by (13). Let ordinary generating functions

$$f(x) = a_0 + a_1 + a_2 x^2 + \ldots, \qquad g(x) = b_0 + b_1 x + b_2 x^2 + \ldots$$

be introduced by

$$f(x) = U\mathscr{A}, \qquad g(x) = U\mathscr{B}$$

where U is the operation used before. Then

$$f(x)g(x) = U(\mathscr{A} * \mathscr{B}). \tag{14}$$

There is a special case when the two sequences coincide: $\mathscr{A} = \mathscr{B}$ so that (14) reduces to

$$f^2(x) = U(\mathscr{A} * \mathscr{A}).$$

Now let (13) be replaced by

$$\binom{n}{0}a_0 b_n + \binom{n}{1}a_1 b_{n-1} + \ldots + \binom{n}{n-1}a_{n-1}b_1 + \binom{n}{n}a_n b_0,$$
$$n = 0,1,2, \ldots. \tag{15}$$

Here we have the binomial convolution sequence: a product $\mathscr{A} \circ \mathscr{B}$ with the nth term given by (15). Let the exponential generating functions be

$$F(x) = a_0 + a_1 x/1! + a_2 x^2/2! + \ldots, \qquad G(x) = b_0 + b_1 x/1! + b_2 x^2/2! + \ldots,$$

that is,

$$F(x) = U_1 \mathscr{A}, \qquad G(x) = U_1 \mathscr{B}.$$

Then (14) is replaced by

$$F(x)G(x) = U_1(\mathscr{A} \circ \mathscr{B}), \tag{16}$$

with the special case

$$F^2(x) = U_1(\mathscr{A} \circ \mathscr{A}).$$

Both the Cauchy and the binomial convolutions extend easily to associative multiplication of any number of sequences. Equations (14) and (16) generalize then to

$$f_1(x)f_2(x) \ldots f_k(x) = U(\mathscr{A}_1 * \mathscr{A}_2 * \ldots * \mathscr{A}_k) \tag{17}$$

and

$$F_1(x)F_2(x) \ldots F_k(x) = U_1(\mathscr{A}_1 \circ \mathscr{A}_2 \circ \ldots \circ \mathscr{A}_k). \tag{18}$$

Since the inverse operations U^{-1} and U_1^{-1} are given by

$$\mathscr{A} = U^{-1}f(x), \qquad \mathscr{A} = U_1^{-1}F(x),$$

both (17) and (18) may be simply written in the bypass form as

$$\pi = VPV^{-1}.$$

In words, this states that the ordinary multiplication π in the generating function domain is conjugate under the operation V (which is U or U_1) to the convolution product P in the sequence domain.

A specific example where only the ordinary generating function works (easily) is provided by the problem of finding the Catalan numbers $C_0, C_1, C_2, \ldots$. These are defined as follows: The initial values are

$$C_0 = 0, \qquad C_1 = 1, \tag{19}$$

and for $n \geqslant 2, C_n$ is the number of complete bracketings of the product

$$X_1 X_2 \ldots X_{n-1} X_n. \tag{20}$$

That is, for the symbols $X_1, \ldots, X_n$ there is defined a binary operation, say multiplication, which is neither associative nor commutative; now C_n is the number of ways of evaluating the product (20). For instance, $C_4 = 5$, since there are exactly five different complete bracketings for $n = 4$:

$$\overline{X_1 X_2 X_3 X_4}, \quad \overline{X_1 X_2 X_3 X_4}, \quad \overline{X_1 X_2 X_3 X_4}, \quad \overline{X_1 X_2 X_3 X_4}, \quad \overline{X_1 X_2 X_3 X_4}.$$

Alternatively, for $n \geqslant 2$, C_n is the number of distinct triangulations of a general convex $(n + 1)$-*gon* P, that is, the number of different decompositions of P into triangles by drawing sides and diagonals so that only the vertices of P may be intersection points.

In the Fibonacci example the recursion was given to us; here it must be produced by examining the structure of the problem. To relate C_n to the C_m with $m < n$, the product (20) may be split as

$$(X_1 X_2 \ldots X_r)(X_{r+1} \ldots X_n), \qquad r = 1, 2, \ldots, n - 1.$$

The first group can be bracketed completely in C_r ways and the second one in C_{n-r} ways; multiplying and summing over r yields.

$$C_n = C_1 C_{n-1} + C_2 C_{n-2} + \ldots + C_{n-1} C_1, \qquad n \geqslant 2.$$

Because of the conveniently chosen initial value $C_0 = 0$ in (19), this recursion may be written as

$$C_n = C_0 C_n + C_1 C_{n-1} + \ldots + C_n C_0, \qquad n \geqslant 2. \tag{21}$$

Now let

$$f(x) = C_0 + C_1 x + C_2 x^2 + \ldots \tag{22}$$

be the ordinary generating function for the Catalan sequence. Since $C_2 = 1$, we have

$$f(x) = x + C_2 x^2 + C_3 x^3 + \ldots + C_n x^n + \ldots,$$

$$f^2(x) = x^2 + \ldots + (C_0 C_n + C_1 C_{n-1} + \ldots + C_n C_0)x^n + \ldots.$$

Therefore by the recursion (21)

$$f^2(x) = f(x) - x.$$

This is a quadratic equation for $f(x)$ and so

$$f(x) = \frac{1 \pm \sqrt{1 - 4x}}{2}, \tag{23}$$

only the negative sign is taken since clearly $C_n \geqslant 0$ for all n. Finally, performing the last part of the bypass, we expand the square root in (23) using the binomial theorem to obtain

$$C_n = \frac{1}{n}\binom{2n-2}{n-1}, \qquad n = 1, 2, \ldots. \tag{24}$$

What would happen if one were misguided enough to introduce, instead of $f(x)$ in (22), the exponential generating function

$$F(x) = C_0 + C_1 x/1! + C_2 x^2/2! + \ldots \quad ?$$

Actually, this question splits into two: What functional relation is satisfied by F? What is the solution F itself? Both questions will be answered, but only by an indirect route: Instead of using the recursion (21) directly, we use our previous knowledge of the ordinary generating function $f(x)$. First, it is found from (22) and (23) that

$$1 - 2F\left(-\frac{x}{4}\right) = 1 + \frac{g(1)}{h(1)}x + \frac{g(1)g(2)}{h(1)h(2)}x^2 + \frac{g(1)g(2)g(3)}{h(1)h(2)h(3)}x^3 + \ldots, \tag{25}$$

where

$$g(z) = \tfrac{3}{2} - z, \qquad h(z) = z^2.$$

Next, we use the following theorem of G. Polya [111, p. 9]: If $y(x)$ denotes the right-hand side of (25) and the following conditions hold:

g and h are co-prime polynomials,
$\deg g \leqslant \deg h, \qquad h(0) = 0,$
$h(k) \neq 0 \qquad$ for $\quad k = 1, 2, \ldots,$

then y satisfies the differential equation whose operatorial form is

$$h\left(x\frac{d}{dx}\right) y = g\left(x\frac{d}{dx}\right)(xy).$$

Applying this theorem we find that

$$xF''(x) + (1 - 4x)F'(x) + 2F(x) = 1.$$

Hence, as is shown in [14, Vol.1, p. 238], $F(x)$ can be expressed by the confluent hypergeometric functions or, equivalently, by the Whittaker functions.

Let us now compare the situations for the Fibonacci sequence $f_0, f_1, f_2, \ldots$ and the Catalan sequence $C_0, C_1, C_2, \ldots$. Both sequences have simple and easily obtained ordinary generating functions but only the Fibonacci sequence has a simple exponential generating function and, in fact, it is something of an accident that the exponential generating function for the Catalan case can be found at all. Why is that so? The answer hinges on the fact that in the Fibonacci case the ordinary generating function is rational while for the Catalan case it is algebraic. It is easily shown that if a sequence has a rational ordinary generating function, then it has a simple exponential generating function as well.

Recall first the definition of an exponential polynomial $E(x)$: It is any expression of the form

$$E(x) = \sum_{j=1}^{s} P_j(x) e^{m_j x}$$

where the P_j are polynomials and the m_j are constants, possibly complex. It will turn out that if the ordinary generating function of a sequence is rational, then its exponential generating function is an exponential polynomial. A simple example is provided already by (5) and (11).

Our assertion is proved simply and directly by using the elementary theory of Hadamard multiplication [134]. If

$$f(x) = \sum_0^\infty a_n x^n \quad \text{and} \quad g(x) = \sum_0^\infty b_n x^n \tag{26}$$

are two power series, then their Hadamard product is the power series

$$f \circ g(x) = \sum_0^\infty a_n b_n x^n$$

with the integral representation

$$f \circ g(z) = \frac{1}{2\pi i} \int_C f(z/w) g(w) \, dw/w, \tag{27}$$

where C is a suitable closed contour containing the origin. Let $g(x) = \exp x$ and let $f(x)$ and $F(x)$ be the ordinary and the exponential generating functions of the same sequence. Then the definition (26) shows that $F = f \circ g$, and the integral representation (27) shows that if f is rational then F is an exponential polynomial.

In conclusion, it may be remarked that the last part of this section shows how interesting mathematical problems arise, even in a simple example, when two equivalent bypasses $W = UTU^{-1} = U_1 T_1 U_1^{-1}$ are compared. However, such technical interest, important though it is, may obscure a central fact already mentioned earlier:

the operations U or U_1 in the bypasses UTU^{-1} and $U_1 T_1 U_1^{-1}$ produce the reflexive power in f: from recursions relating some members of a sequence to other members we obtain, after U or U_1, a relation of f to itself.

This relationally reflexive power deserves a special note. The present example involving generating functions is about a simple technical tool from the arsenal of almost every practicing mathematician. The next example will concern a deep and subtle foundational topic, Gödel's incompleteness theorem, which might appear totally different in every respect. But the juxtaposition of generating functions with Gödel's theorem is not accidental in spite of the title of the present chapter. The heart of Gödel's proof is a subtle construction of a proposition G that asserts its own unprovability. That is, Gödel succeeds in legitimately endowing his proposition G with the relationally reflexive power of referring to itself. Just how is this bootstrapping done? Not surprisingly, we shall claim: by means of a bypass. So, perhaps the topics of this example and the following one are not so far apart as they might appear. Or, less modestly put, the diameter of the field to which bypasses apply is at least as big as the distance separating generating functions from Gödel's theorem.

EX. 7.2. GÖDEL'S FIRST INCOMPLETENESS THEOREM

This theorem is considered to be one of the greatest intellectual achievements of our century or indeed of all time. First exposition to it may lead to an uneasy impression that something must be wrong somewhere because here is someone who appears to claim success in lifting himself by pulling on his own bootstrap. Later, one gets used to Gödel's argument and one comes across some of Gödel theorem's junior relatives, such as certain results on the universal Turing machines [37] or the self-replicating automata of von Neumann [105]. Increasing familiarity may breed contempt, even for the fearful symmetry of Gödel's tiger, which might be judged more excellent than that of Blake's.

Let us leave out the important though purely technical distinction between consistency and ω-consistency, and let us delay till later a discussion on the importance of mathematical and metamathematical levels [80]. Then Gödel's theorem is simply stated: If ordinary arithmetic is consistent, then it contains undecidable statements. These are statements whose truth or falsity cannot be proved within arithmetic.

The essential tool for the proof is the device of Gödel numbering. Since the alphabet of arithmetic is countable, it is possible to produce a method of ordering, or numbering, that assigns a unique positive integer to every correctly framed arithmetic formula. The same is true about strings of such formulas: A unique positive integer is assigned to every finite string of such formulas separated, say, by commas. There are many such numberings; in some every positive integer is the G.N. (Gödel number) of a formula or a string of formulas. In others only certain special numbers may occur as G.N.'s. It is irrelevant what particular numbering is used; all that matters is that a numbering be fixed under which different formulas or strings have different G.N.'s. Let it be supposed that a specific numbering is used in which all formulas and strings are ordered in the form of the natural number sequence (i.e., first, second, third, etc.). In particular, a proof of a formula F is a finite string starting with the axioms and ending up with F; thus proofs also receive G.N.'s in the ordering.

The pivot of Gödel's proof [57] is a formula G that asserts, though only on a suitable interpretation, its own unprovability; it is constructed as follows. Let $P(a,b)$ be the statement

    a is the G.N. of a formula whose proof has G.N. b,

and consider the statement

$$(b) \qquad \bar{P}(a,b) \tag{1}$$

where (b) ... is the universal quantifier that reads: for all b..., and the bar denotes negation. Then (1) reads: No number b is the G.N. of a proof of the formula whose G.N. is a, or, briefly, the ath formula in our ordering is unprovable. There is in (1) exactly one free, that is, uncommitted, variable: a. This one degree of freedom is used to perform the diagonal process of Cantor. Namely, Gödel shows, though this takes him more than half his long paper [57], that the formula (1) itself is Gödel-numberable; let its G.N. be q. That is, the formula (1) with one free variable a is identical with a specific formula, the qth one in the ordering used. Cantor diagonalization means substituting the definite value q for the free variable a in (1). This reflex bending yields the undecidable formula G, which is simply

$$(b) \qquad \bar{P}(q,b)$$

asserting that it itself is unprovable.

Now, how does the bypass principle enter here? As remarked by Gödel himself in [57], there is an obvious analogy between his undecidable G and such classical paradoxes as the Liar paradox or the Berry paradox [80]. The Liar paradox is simply: The statement I am making here is false. The Berry paradox concerns "the least natural number not nameable in fewer than twelve words"; it succeeds in naming with eleven words a natural number that by its original definition cannot be named with fewer than twelve words. As is seen, these paradoxes depend for their existence on the device of self-reference. For future use, we may call such types of self-reference gross or direct.

The essence of Gödel's argument is that his construction of the undecidable formula G avoids the use of such gross or direct self-reference. Instead, the self-reference, or relationally reflexive effect, is achieved (as we shall argue) by means of a bypass. The result is a legitimate statement and not a paradox.

We quote now from Gödel's paper [57, footnote 15] a sentence referring to the undecidable formula, or theorem, G: "Such a theorem has in spite of contrary appearance nothing circular in itself since it asserts, to begin with, the unprovability of a quite definite formula (namely, the qth one in the lexicographical ordering, on a definite substitution), and only later it develops (so to speak, by accident) that this formula is just the one expressing the theorem itself."

All this is summed up in the following bypass schema giving the schedule of operations used in constructing the undecidable proposition G:

Start with the formula $(b)\bar{P}(a,b)$ to get G

then use the G.N. encoding   finally use the
                                                           G.N. decoding

                                        then diagonalize
to obtain the G.N.   to obtain
of the above q ———————→———— $(b)\bar{P}(q,b)$

and interpret this G as the formula asserting its own unprovability. To put it more starkly, but perhaps more suggestively, the formulation (C_2) from Chapter 1 might be employed. The forbidden gross or direct self-reference represents the wall W. Gödel gets past it and produces, legitimately, fine or induced or indirect self-reference. This he accomplishes with the device of conjugating diagonalization by means of the Gödel numbering. This view of Gödel's theorem could be supported by the previously quoted passage from Gödel. He says there: "...and only later it develops...." What does "later" refer to? Presumably, to the juncture when the third and last stage of the process is finished, when the Gödel numbering has been used "in reverse" and exactly for the purpose of interpreting the formula G as one that asserts its own unprovability.

The preceding bypass schema for Gödel's theorem suggests a considerable analogy to the process of translation. It is as though, faced with a problem stated in one language, we were to translate it and work on it in another language, and finally retranslate back to the original language, where the result is verified to be the solution. Here a comment may be made in reference to the mathematical and the metamathematical levels mentioned earlier. Roughly speaking, in the context of this whole example the mathematical level refers to (general) statements about numbers, and the metamathematical level refers to (general) statements about (general) statements about numbers. Or the one refers to theorems, and the other to theorems about theorems. The device of Gödel numbering performs a translation service: It enables us to pass from the one level to the other by exactly mirroring above whatever happens below, and conversely. So, one starts at the metamathematical level, passes from there to the mathematical level, where the required diagonalization is performed thus achieving a closure, and then one returns to the meta-mathematical level, where the proposition obtained is interpreted as referring to itself.

As is well known [80], in the narrower system of arithmetic with addition alone but without multiplication, such exact mirroring does not occur. Here we have the theorem of Presburger: The narrower system is decidable. That is, every proposition in it is either provable or disprovable.

A relation was claimed earlier between Gödel's theorem, on the one side, and the universal Turing machines as well as von Neumann's self-replicating automata, on the other. The relationship is not hard to show: All three of them operate on a bypass that makes effective self-reference conjugate to an infinite closing operation which is the Cantor diagonal process. A very much cruder but still similar process occurs already in connection with normal functioning of the tank, in the fourth preliminary example of Chapter 1. The tank rolls itself along the iron bridge of its cleats, and this bridge is closed up and so, in effect, endless. What the Gödel numbers really provide is a language, one that operates between the two levels. It might be argued, and we shall attempt it, that ordinary language too is a bypass; that it, too, operates between a "level" and a "metalevel," and finally that it, too, makes effective self-reference possible.

One last note is needed, this concerns the Paris–Harrington theorem (PHT) from Chapter 2. It will be remembered that the PHT, like Gödel's proposition, is an example of a statement that is true but unprovable (within the ordinary Peano arithmetic). Yet there is a great difference between the two. It might be objected that Gödel's proposition is not "about" anything (unless it be about itself). That is, it does not concern a *previously* formulated mathematical situation but, rather, has been produced so as to lead to unprovability. On the other hand, the PHT grew out of combinatorics, and so may be said to have arisen naturally. Or, the one has been manufactured, the other has been born. It is therefore of interest that the PHT has been proved in [13] by starting from its combinatorial context and mirroring within it a Gödel-like situation: A model T for the Peano arithmetic is set up within which the combinatorial principle of the PHT implies the consistency of T.

EX. 7.3.   SETS, ORDINALS AND CARDINALS, PATTERNS

The founder of set theory was a German and so were many of the early workers in the field; their works often start with some such phrase as that of Hausdorff

[65, p. 8]: *"Eine Menge ist eine Vielheit, als Einheit gedacht"*—"a set is a multiplicity thought of as unity." In the later development, and even in the work of the pioneers themselves (except for the founder, Georg Cantor), the psychological element implicit in that phrase was soon given up. In a way, this is a pity because parts of set theory might be a useful tool to psychologists, sociologists, and others. To suggest this, let us turn the preceding phrase of Hausdorff inside out: Thinking may be usefully analyzed in terms of the operation of collecting many into one and of the inverse operation of separating one into many.

It makes neither much imagination nor particular partisanship to observe that those two reciprocal operations might form the outer parts $S \ldots S^{-1}$ of some important bypasses STS^{-1}. If set theory is indeed to be useful in the biosociological fields, then such usefulness can come best from the workers in those fields themselves. Unfortunately, the standard mathematical treatments of set theory are sufficiently forbiddingly technical to put off non-mathematicians (and often mathematicians as well). In particular, it may appear from such treatments that sets and certain operations on sets (which lead to the fundamental concepts of ordinal and cardinal numbers) are something immensely abstract. The exact contrary is believed by the author; the odd exposition that follows is an attempt to substantiate that belief, and not only to suggest the usefulness of bypass concept in this direction.

The situations to be considered here are conveniently introduced by the following illustration. In a certain hypothetical country, which might be thought of as an idealized Scotland, there is a simple method of social grouping based on clans. There is a number of such clans and everyone belongs to just one clan; all members of a clan have the same family name, which nobody else has; finally, each clan has exactly one chief. Then each Scot may be characterized in three different though related ways: by name, by clan, by clan chief. Thus someone named Ian Macdonald is characterized in the first way by his family name, which is "Macdonald"; in the second way by his clan which is "the Clan Macdonald"; and in the third way by the chief of his clan who is "The Macdonald." When referring to the chief, the first letter of the definite article is always capitalized and its last letter is always accented, so that the chief is distinguished in both written and spoken language. This use of the definite article to single one out of many is of course based on Scottish actuality; it is a striking device and has indeed suggested the present illustration.

The main point is that something quite different gets attached to our individual Ian Macdonald in each of those three ways: first a name, which is an abstract label or index, then a certain collection of people, in which he himself is included, and finally a certain special individual of that collection, who might even be Ian Macdonald himself if he happens to be the chief of his clan. The three attachments are not on a par: The name and the clan are mostly, though not always, interchangeable; the chief usually is not interchangeable with either. This disparity becomes clearer if our Scotland is taken to be nameless, clanless, and chiefless, but is provided with a kinship relation K. That is, every two Scots x and y either are kin or are not: xKy or $x\bar{K}y$. Kinship is an equivalence relation since it is taken to be

reflexive	(xKx for every x),
symmetric	($xKy \leftrightarrow yKx$, for every x and y),
transitive	(if xKy and yKz then xKz, for every x, y, and z).

Therefore the whole population divides into disjoint equivalence classes, or kinship classes, and two individuals are kin if and only if they belong to the same class. The equivalence class is the clan and belonging to it is signified by the label of the name. So it is tempting to say that the name is as good as the clan. But kinship alone does not determine the chiefs, even though it determines both clans and names. Some rule is needed by which the chiefs are picked, chosen, elected, or perhaps elect themselves.

Serious possibilities for bypass emerge already, even at this primitive stage of illustration. Consider our Ian Macdonald and a member of another clan, say Flora Douglas. While certain relations between them are private and concern no one else, certain others, such as disagreements, disputes, or dishonor, may be considered public so as to concern their entire two clans, and in particular their two clan chiefs. This may lead to a fight between the clan Douglas and the clan Macdonald, or perhaps to an adjudication between The Douglas and The Macdonald. In the latter case the relation of Ian and Flora is symbolized by the compound or stacked bypass

$$\phi R A R^{-1} \phi^{-1}, \tag{1}$$

where ϕ stands for belonging to and ϕ^{-1} for including, R for being represented by and R^{-1} for representing, and A for adjudicating or, rather, for adjudicating with. The skeletal form (1) of the bypass needs to be filled in with the proper names of clans and individuals, and it becomes then

$$\text{(I.M.)}\phi\text{(Cl.M.)}R\text{(T.M.)}A\text{(T.D.)}R^{-1}\text{(Cl.D)}\phi^{-1}\text{(F.D.)}. \tag{2}$$

In words this is: Ian Macdonald belongs to the Clan Macdonald, which is represented by The Macdonald, who adjudicates (the case together) with The Douglas, who represents the Clan Douglas, which includes Flora Douglas.

Some remarks are in order here. Attention may be drawn to the grammatical peculiarity of the active and the passive forms appearing as inverses. This extends even to the relation ϕ which is just the mathematical $\in$ as in $x \in X$: The individual x belongs to the collection X. The inverse $\in^{-1}$ is not used in mathematics and, curiously enough, "belongs to" has no passive form in ordinary language either. However, both would have referred to inclusion, as in $X \in^{-1} x$: The collection X includes the individual x. Here the verb "include" is used rather than the verb "contain"; the former goes with the membership relation and the latter is reserved for the subset relation. Thus, Ian Macdonald is included in the Clan Macdonald, whereas he and his family form a subcollection contained in the clan. This remains so even if the whole family consists of Ian alone. Finally, it may be remarked that (1) and (2) stand somewhat in the relation of propositional function and proposition; the matter of bypasses forming certain symmetric propositional functions may be of some interest in logic.

To return to bypasses, we note that another clan—individual—clan bypass might arise in a collective disagreement between the two clans which results in a trial by combat between the two chiefs:

$$RFR^{-1} \tag{3}$$

or, with proper insertions,

$$(Cl.M.)R(T.M.)F(T.D.)R^{-1}(Cl.D.). \tag{4}$$

This reads in brief: The Macdonalds are represented by The Macdonald, who fights The Douglas representing the Douglases. It may happen that the chiefs are allowed to choose champions, and The Macdonald chooses Ian Macdonald while The Douglas chooses John Douglas. Then there appears a new pair of individual–individual relations C and C^{-1}: C stands for being championed by, and C^{-1} for championing. The bypass (4) is now stacked in depth to

$$(Cl.M.)R(T.M.)C(I.M.)F(J.D.)C^{-1}(T.D.)R^{-1}(Cl.D.), \tag{5}$$

or in its skeletal form to: $RCFC^{-1}R^{-1}$.

It will be clear by now that our lengthy illustration is meant, among other things, to introduce a type of bypass that may be called i/c bypass (where i/c is the mnemonic for individual/collective). The idea is that relations between two individuals may be, or may have to be, mediated through certain two collections to which the two individuals belong. Conversely, relations between two collections may be, or may have to be, mediated through some two individuals, or some special two individuals, representing those two collections.

The principal i/c bypass seems to be that one which assigns to an individual of a collection some unique individual of that collection, as in the instance where The Macdonald is assigned as clan chief to our clansman Ian Macdonald:

$$
\begin{array}{ccc}
\text{Individual} & \text{Ian Macdonald} & \text{The Macdonald} \\
& \phi \downarrow & \phi^{-1} \uparrow \\
\text{Collective} & \text{Clan Macdonald} \xrightarrow{\ T\ } & \text{Clan Macdonald}
\end{array}
$$

Here ϕ and ϕ^{-1} are as before and T stands for selecting, or for the method of finding, or for determining who is, the chief within the clan. It will be shown that for suitable T the preceding bypass serves with very little change to define first the ordinal numbers, then the cardinal numbers, and finally that elusive concept known as pattern (or as concept).

Behind all this there looms a philosophical puzzle, or problem, or pseudo-problem. In its ancient appellation it is usually known as the problem of the One and the Many. In its medieval guise, it turns up in the nominalist-realist controversy of the scholastic philosophy about the status of individual and of collective entities. In its modern form it reappears in the dispute between intuitionism or more rabid forms of constructivism, and the less radical schools of mathematical logic and foundations of mathematics. Yet, in one form or another, and elusive though it might appear, this puzzle or problem or pseudo-problem hovers around the foundations of physics, biology, psychology, sociology, and politics (to name just a few subjects). The inclusion here of such an eminently practical domain as politics may be surprising. But it will be remembered how the nominalist-realist controversy soon acquired a political twist in the central historical happening of the Middle Ages: the contest between the Emperor and the Pope. The medieval nominalists, with their insistence on the importance of the individual, represented intellectually the "left wing" of their day and were mainly on the lay side. The realists, who emphasized the importance of collections (species, genera), were the "right-wing" counterpart, corresponded to what today would be called the idealist school, and tended to side with religion. There were also some isolated attempts to resolve the problem by going beyond the dichotomy.

The most eminent effort to this end was that of Peter Abelard, who was first a student of the outstanding nominalist Roscellinus, and then of William of Champeaux, the champion of realism. The direction of Abelard's mature thought is indicated by the title of his most famous book *Sic et Non*, which means "Yes and No." This book dealt with theological controversy and its main point was that argument by appeal to authority was vain since for any question authorities could be cited on either side (hence "Yes and No"). But it is tempting, and not entirely inappropriate, to observe that the "Yes and No" of Abelard also sums up his view on the whole subject of the nominalist-realist controversy: yes—the realists and the nominalists are both partially right, no—neither side has the whole truth exclusively. Or, in terms of issues rather than of schools, both the individuals and the collections are real and useful—within limits; neither is exclusively so. By stretching the point, Abelard might be regarded as the forerunner of Niels Bohr in respect to the latter's famous complementarity principle [59, p. 3]. This is an attempt to resolve the particle-wave duality in physics, and a part of the principle asserts that in the atomic domain both concepts, that of particle and that of wave, are necessary for an adequate description of phenomena. Abelard tried to achieve a similar resolution of the individual-collective duality and, in fact, his title *Sic et Non* is the briefest possible statement of a complementarity. Eventually it may well turn out that Abelard's complementarity and Bohr's are not quite as different as they may at first seem to be.

The relations between the one and the many are quite obviously important in such practical subjects as statistics and probability, and here again the i/c bypass turns up as a conceptual framework. For a simple example, consider the statement that the Macdonalds are on average 5 ft. 10 in. tall. Just how does this numerical label, or statistic, get attached to an individual Macdonald, for instance, to our Ian who is 6 ft. 4 in. tall? There is here an i/c bypass $\phi M \phi^{-1}$ that starts with Ian, makes the passage ϕ to the clan, follows this by the process M of measuring all clan members (or perhaps a sample of them) as well as counting them and making a simple calculation of addition and division, and the resulting measurement 5 feet 10 inches gets attached to Ian when we return to him by the last part ϕ^{-1} of this bypass. Note that the mean might be ascribed to the whole clan rather than to its members; on the other hand, we are often interested in relative judgments, such as how tall a particular Macdonald is relative to his clan. This point, together with the last part ϕ^{-1} of the bypass, is clearer when our Ian is assigned to some quartile, decile, or percentile or when we locate his height at 1.8 standard deviations above the mean.

The symbol ϕ stands for the relationship of being a member of; in using sampling to compute the mean we really employ something like a stacked bypass whose skeletal form is $\phi \psi M \psi^{-1} \phi^{-1}$. Here ψ is the relationship of containing a subset, so that the preceding stacked bypass, after being properly filled in, becomes

$$(I.M.)\phi(Cl.M)\psi(Sample)M(Sample)\psi^{-1}(Cl.M)\phi^{-1}(I.M.).$$

In words this reads: Ian Macdonald is a member of the Clan Macdonald which contains a subcollection (i.e., the sample) on which the operation M is carried out giving the sample mean, which is transferred to the containing set of the whole Clan Macdonald and eventually attached to the included member of the set, Ian Macdonald. Of course, here we come again across the very important question: What is lost (and what is gained) in such a bypass?

It is sometimes said that our sole means of procuring new scientific knowledge is by the process of inductive reasoning or, more precisely, by that process as an integral part of the scientific method, thus leading to meaningful prediction. A possible i/c bypass is hidden here by its obviousness: Presumably, the start is with observations on the *individual* level, this is followed by formulation and proving of hypotheses, proofs, and theorems on the *collective* level, and finally these are used to predict the behavior of *individuals*. In some cases, as for instance with the natural sciences, this hypothesized bypass is not simple but stacked, so that its middle part, that which occurs at the collective level, is another bypass: from the physical level of natural phenomena by a process of translation into the mathematical level of quantities and symbols, working there, then reversing that translation process to interpret the results again in terms of natural phenomena. Hence there arises a question which is perhaps best formulated in the title of a paper by E. Wigner: "On the Unreasonable Effectiveness of Mathematics." Simply: Why does it work? An answer given by Plato is, briefly: The creator worked with numbers as a blueprint, hence the creation reflects its blueprint, and so numbers work. Nowadays such an answer may be no longer (or perhaps not yet) acceptable. To say that mathematics works on account of bypasses is not much of an answer; it is a little better to answer that mathematics works because of bypasses and the structures preserved and introduced through bypasses. It is best, in our context, to leave the matter with a suggestion that is really a repetition of some earlier statements. The claim was that conjugacy is a general method of resolving complexities into simpler components and that it is a part of nature's and mind's software. That, on suitable interpretation, may be a part of the reason why "it works."

One final comment will be made before moving on to ordinal and cardinal numbers. We have used, and shall use, the simple sociology and anthropology of our idealized Scotland to illustrate certain relations of sets and individuals. In particular, we have used our illustration to introduce the i/c bypass easily and conveniently. On account of the anthropological flavor of our terms, one might ask whether genuinely anthropological examples of i/c bypasses exist.

We submit now a candidate; this concerns what A. van Gennep [53] has called a rite of passage and is based on the classic work of N. D. Fustel de Coulanges [32, Bk.2, Ch.2] on the ancient Greek and Roman marriage ritual:

The Roman marriage closely resembled that of Greece, and, like it, comprised three acts—*traditio, deductio in domum, confarreatio.*

1. The young girl quits the paternal hearth. As she is not attached to this hearth by her own right, but through the father of the family, the authority of the father only can detach her from it. The *traditio* is, therefore, an indispensable ceremony.

2. The young girl is conducted to the house of the husband. As in Greece, she is veiled. She wears a crown, and a nuptial torch precedes the cortege. Those about her sing an ancient religious hymn. The words of this hymn changed, doubtless with time, accommodating themselves to the variations of belief, or to those of the language; but the sacramental refrain continued from age to age without change. It was the word *Talassie*, a word whose sense the Romans of Horace's time no more understood than the Greeks understood the word ὑμεναιε, and which was, probably, the sacred and inviolable remains of an ancient formula.

The cortege stops before the house of the husband. There the bride is presented with fire and water. The fire is the emblem of the domestic divinity; the water is the lustral water, that serves the family for all religious acts. To introduce the bride into the house, violence must be pretended, as in Greece. The husband must take her in his arms, and carry her over the sill, without allowing her feet to touch it.

3. The bride is then led before the hearth, where the Penates, and all the domestic gods, and the images of the ancestors, are grouped around the sacred fire. As in Greece, the husband and wife offer a sacrifice, pouring out a libation, pronouncing prayers, and eating a cake of whaten flour (*panis farreus*).[7]

Footnote (7) reads:

We shall speak presently of other forms of marriage in use among the Romans, in which religion had no part. Let it suffice to say here, that the sacred marriage appears to us to be the oldest; for it corresponds to the most ancient beliefs, and disappeared only as those beliefs died out.

Some preliminary remarks are in order.

1. The domestic hearth, which figures so prominently in the preceding passage, was called in Latin *focus*, a mathematically interesting word.

2. The first part of the Roman marriage ceremony, called *traditio*, the separation of the bride from the paternal hearth and authority, survives in the ritual question of the Anglican wedding ceremony: Who giveth this woman in marriage?

3. The carrying of the bride across the sill need not signify any symbolic violence. It may well be a ritual cutting off, or separating, associated with the magical beliefs attached to the threshold or sill. Similar ritual beliefs, among the twelfth- and thirteenth-century Mongols, are re-

corded by Father da Plano Carpini in his *Tartar Relation* [129], and in other places for other peoples. The whole matter may even reach very far back to human beginnings: The sill might symbolize the entrance to the cave.

4.   The sacred Greek refrain ὑμέναιε, of course, gives rise to the English hymeneal, which is a synonym for marriage.

5.   The veil of the bride as well as the nuptial torch carried before her may underscore the "rite of passage." The bride has been separated from one worship, but not united with another worship. Therefore, she is temporarily outside the pale and particularly liable to malevolent demonic influence; so, she must be protected from it.

We come now to our point. Then, as now, the female figures in the ceremony far more prominently than the male; it is she who makes the crossing, not he; so we speak of "husband and wife" but of "bride and groom." The whole passage quoted earlier is used by Fustel de Coulanges to support his claim that the earliest Greek and Roman beliefs and rites were based on a complete identification of social group with religious congregation. Both collections were one and the same: extended family. It follows that the act and fact of membership were taken all the more seriously. Hence comes the i/c bypass character of the described ancient marriage ritual; all three bypass parts of it are important and are clearly marked in the quoted passage: An individual is detached from one collection, conducted to another, and then she is attached to the other collection.

Our attention turns now to the ordinal and cardinal numbers. For the finite case no distinction need be made between ordinals and cardinals, but we want a uniform treatment of both finite and infinite numbers. The ordinals and cardinals arose historically from the need to compare infinite as well as finite sets with respect to order and with respect to size. They are the single-handed creation of Georg Cantor (1845–1918) [26], whose work was later refashioned and formalized by Zermelo, Fraenkel, Russell, von Neumann, Bernays, Gödel, and others (see Reference 15). The ordinal numbers came first. Like some other mathematical fundamentals, such as logarithms or Fourier series, they had a difficult birth. It was only in their later development that they acquired such simplicity and naturalness as they might now be said to possess.

In his work on trigonometric series [26], Cantor used in 1872 the notion of the derived, or limit, set S' of a set S of real numbers. He was led to iterate this priming operation; at the first the iteration went to the second, third, ..., nth derived set S'', S''', ..., S^n, ... Then the iteration was pushed into the transfinite domain, to the ωth, $(\omega + 1)$st, ..., αth derived set S^ω, $S^{\omega+1}$, ..., S^α, ... Here ω is the first infinite ordinal, $\omega + 1$ is the next one, and so on. This shows that the ordinal numbers arose from the need of indexing very long sequences, that is, transfinite sequences of arbitrarily high order of infinity. Briefly, ordinals are what sequences get indexed with. This shows itself even in the matter of names: "sequence" comes from Latin *sequor*, "I follow"; this refers to rule or order, hence "ordinal." Another point to emphasize is the role of iteration, a topic we have already met in Ex. 5.7.

There are three different major ways of introducing ordinals. The original method is due to Cantor himself, this was followed by Russell's method which, in turn, was superseded by von Neumann's method. In terms of our Scottish illustration these three methods can be sketched very simply: Cantor defines an ordinal by its family name, Russell by its clan, and von Neumann by its clan chief. So, roughly speaking, the difficulties facing the three methods are as follows.

Cantor needs to produce the abstract label which is the name; it is quite unimportant that he actually uses a label called "order type" rather than "name." On account of that abstract label Cantor's definition is sometimes called, as in Reference 15, definition by abstraction. So, what Cantor needs is a "family name act," and his definition is objectionable on the grounds of parsimony: Extraneous objects (names, or order types) are introduced.

This objection is avoided by Russell, who defined an ordinal as the clan of all objects that are, in a suitable sense, "kin" to each other. Again, it is quite irrelevant that Russell's terminology uses "class" rather than "clan." Russell's difficulty is that of delimitation: Class turns out to be embarrassingly huge; so, what he needs is a "citizenship act" or "membership act."

Loosely speaking it might be said that Cantor is forced to work outside the clan and Russell to work on its boundary. Unlike them, von Neumann works inside the clan and produces a rule for picking out a unique individual, the clan chief, which is defined as the ordinal number. Hence von Neumann's theory is apt to be more complicated because of its need for the rule of picking out unique chiefs of clans. In a very much simpler context this sort of problem was already encountered in Ex.4.0, with the circle C, ellipses E, rectangles R, and the square S. It may be worth noting that the bypass of that simple problem also concerned set membership and, in fact, was what we call an i/c bypass. Further, it may be observed that the circle C is distinguished among ellipses E by its particular symmetry. As we shall see later, von Neumann also distinguishes the unique clan chief, which is the-ordinal-number-to-be, by its particularly symmetric structure. Another small but interesting point here is a recurrence of the already mentioned grammatical peculiarity of the definite article: We spoke of the circle C (among ellipses E) and of The Macdonald (among Macdonalds).

After the preceding preliminaries we take up the ordinals themselves. Since they serve for indexing, or ordering, it is necessary to examine this concept first. Consider the following five sequences:

$$x_0, x_1, \ldots, x_n,$$
$$x_0, x_1, \ldots, x_n, \ldots,$$
$$\ldots, x_n, x_{n-1}, \ldots, x_1, x_0,$$
$$x_0, x_1, \ldots, x_n, \ldots, x_\omega,$$
$$x_0, x_1, \ldots, x_n, \ldots, x_\omega, x_{\omega+1}, \ldots, x_{2\omega},$$

or, what is really of sole interest, the corresponding sequences of indices themselves:

$$0, 1, \ldots, n,$$
$$0, 1, \ldots, n, \ldots,$$
$$\ldots, n, n-1, \ldots, 1, 0 \qquad (6)$$
$$0, 1, \ldots, n, \ldots, \omega,$$
$$0, 1, \ldots, n, \ldots, \omega, \omega + 1, \ldots, 2\omega$$

The first sequence is finite, it has a first and a last member, and it happens to be a properly initial segment of the next sequence. This next sequence is the standard infinite sequence of all natural numbers in their standard order; it has a first, but not a last, member. The third sequence is the second one reversed; it also consists of all natural numbers but written in order that is the reverse of the usual; there is a last member but not a first one. The fourth sequence is transfinite: It is an infinite sequence of which the second one is a properly initial segment. There is a first member 0, but also a last member ω. This ω is, or rather will be, the first infinite ordinal, and it is preceded by all the natural numbers which, of course, are the usual finite ordinals. Therefore, ω, being the first infinite ordinal, has no predecessor: It is an example of what the theory of ordinals calls a limit ordinal. The last sequence in (6) is also transfinite, and it consists of two copies of the preceding sequence

$$0, 1, \ldots, n, \ldots, \omega; \qquad 0', 1', \ldots, n', \ldots, \omega'$$

if we write $\omega + n$ for n' so that the last member ω of the first copy and the first member $0'$ of the second copy are merged into single element. This last sequence contains two limit ordinals, ω and 2ω.

Every sequence in (6) is more than a mere heap of its elements because these are ordered in the following sense: In each case there is a rule of precedence under which

(1)   of any two distinct elements one must precede the other, and

(2)   if x precedes y and y precedes z, then x precedes z.

Any collection (i.e., heap) of elements is called *ordered* if there is defined for it a rule of precedence that satisfies (1) and (2). We should like to exclude somehow the third sequence in (6) on account of its anomaly of not having anything to start with: Infinite regress bothers us more than infinite progression. One might even claim that to call the third member of (6) a sequence is an abuse of language. The same exclusion should apply to the reversal of the fourth sequence in (6):

$$\omega, \ldots, n, n-1, \ldots, 1, 0.$$

Here we do have a first element but if this first element is left out we get again the "beginning-less" third sequence of (6). On the other hand, if we take the first, second, fourth, or fifth sequence in (6) and remove any part of it, then what remains will always have an initial element. These remarks motivate the following definition: Any collection X is called *well-ordered* if it is ordered by a rule of precedence which satisfies (1) and (2) above, and also the further requirement

(3) If Y is all of X or any subcollection of X, then Y contains a unique element that comes first in the order of precedence.

The well-ordered sets are the material to define ordinal numbers on. Since the ordinals are to index sequences with, we do not want to distinguish between well-ordered sets that do the same indexing job, for instance, between

$$0, 1, 2, 3, \ldots, n \qquad \text{and} \qquad 0, 2, 4, 6, \ldots, 2n$$

or between

$$0, 1, 2, 3, 4, \ldots \qquad \text{and} \qquad 0, 1, 4, 9, 16, \ldots.$$

Therefore the relationship of similarity is introduced: Two well-ordered sets X and Y are similar if and only if there is a 1:1 correspondence between their elements, which preserves precedence. That is, if x precedes y in X and x corresponds to u while y corresponds to v, then u precedes v in Y, and vice-versa. The individuals making up the population of our country are now the well-ordered sets and similarity constitutes the kinship. Therefore family names and clans arise, as was shown. So, we have the Cantorian definition: An ordinal number is an order type of a well-ordered set, and the Russellian definition: An ordinal number is an equivalence class in the collection of well-ordered sets under the similarity relation. This is exactly what was meant by a previous statement in terms of our illustration: Cantor defines an ordinal number as "the name," and Russell defines it as "the clan."

Von Neumann prefers to define an ordinal number not as the whole equivalence class, like Russell, but as a unique canonical member of it. This is exactly what was meant by saying that ordinals are now defined as clan chiefs. To pass on to details, let us first recall von Neumann's construction of ordinals out of sets, which was briefly mentioned in Chapter 2. The empty set $\varnothing$ is defined as the initial ordinal 0, 1 is defined to be the (one-element) set $\{\varnothing\}$ or $\{0\}$, 2 is defined as the (two-element) set $\{\varnothing,\{\varnothing\}\}$ or, what is the same thing, as $\{0,1\}$, and so on. Thus each ordinal after 0 is exactly the set of all its predecessors. As such, it has a remarkable property: every element of it is also a subset of it. Or, equivalently, the $\in$-relation is transitive on ordinals:

$$\text{if } x \text{ is an ordinal, } y \in x, \text{ and } z \in y, \text{ then } z \in x.$$

Sets with this property (that a member of a member is itself a member) are called *transitive*. Now, the von Neumann definition is:

$$\text{an ordinal is a transitive set well-ordered by } \in.$$

As it turns out, one can prove under a very mild assumption that any well-ordered set is similar to just one ordinal: Each clan has exactly one chief. Thus $\{0,1\},\{0,2\},\{1,2\},\{3,\omega\},\ldots$ are all members of the "clan of pairs" and are all well-ordered by $\in$, but only the first one of them is transitive; it is therefore the clan chief or, as we usually call it, the ordinal number 2.

The situation with the cardinal numbers is as follows: They measure solely the size or numerosity of sets. Thus, to compare two sets ordinally is to decide which one is a longer sequence, to compare them cardinally is to decide which one is a bigger heap. Some remarks of Cantor himself are in order here. Given a (well-ordered) set X, Cantor denotes by $\bar{X}$ its order type and he calls the putting of the bar on X the first abstraction act: to disregard wholly the structure and nature of the elements of X, except for their order or ranking. One might add that in placing the bar on X we homogenize its members while retaining their order. Further, Cantor denotes by $\bar{\bar{X}}$ the power, or cardinality (i.e., mere size) of X, and he calls the placing of the two bars on X the second abstraction act: to disregard not only the structure and nature of the elements of X but also their order. That is, now we not only homogenize its members, but we also mix them.

As before with the ordinals, we do not want to distinguish between the cardinal numbers that perform the same job of size measurement. To make this precise, we say that two sets have the same cardinality (or power, or size) if there exists a 1:1 correspondence between their elements. For instance, the sets of integers and of all algebraic numbers have the same cardinality, as do the sets of all real and all complex numbers. Further, a set X is smaller than a set Y if X can be placed in a 1:1 correspondence with a subset of Y, but Y cannot be placed in a 1:1 correspondence with a subset of X. Since we already have the ordinal numbers to work with (because they have been defined already), we define the cardinals in terms of ordinals: a cardinal number is any ordinal bigger than its predecessors. Hence $0,1,2,\ldots,\omega$ are all cardinals. However,

$$\omega + 1, \ldots, 2\omega, \ldots, \omega^2, \ldots, \omega^{\omega^{\cdot^{\cdot^{\cdot}}}} \quad (\omega \text{ times}), \ldots$$

are all "equally big"—just as big as ω or, in the standard terminology, they are all countable ordinals; none of them is a cardinal.

We come now to our last topic: patterns. The difficulty of dealing with "pattern" is underlined by the mere existence of the many nearer and further relatives of that difficult word: shape, form, configuration, character, gestalt, idea, archetype, concept, and so on. The interest here is in two questions: What is a pattern? How are patterns distinguished? Instead of the act of distinguishing one could speak of detecting, recognizing, discriminating, perceiving, or abstracting. Our treatment of patterns is borrowed from [96, pp. 355–370].

To start with an example, we can distinguish a triangle from a quadrilateral and either of them from a pentagon. Now, what does it mean to say that an n-gon can be distinguished as such, no matter what its size, angles, or position may be? In its simple idealized version at least, the problem of patterns is the problem of set membership for disjoint sets. Thus, in our example of n-gons, we have a collection $P_3, P_4, \ldots$ of sets, with the indices $3, 4 \ldots$; P_n is the set of all polygons with n sides (or n vertices), and the number n is the "pattern."

Generally, let there be given a collection

$$\{X_i : i \in I, \, X_i \cap X_j = \emptyset \quad \text{if} \quad i \neq j\}$$

of pairwise disjoint sets indexed by some set I, and let x be a member of one of the sets:

$$x \in \bigcup_{i \in I} X_i.$$

Then there is a unique index depending on x, $i = i(x)$, such that x is in the set X_i. The index set I is the set of patterns, and in particular, the index $i = i(x)$ is the pattern of x.

All this may be expressed simply, but reasonably precisely, in terms of our Scottish illustration. To be able to identify patterns means this: On being presented with any Scot to be able to identify the clan to which that Scot belongs. It does not matter whether the name, the clan, or the chief (or something else) is used for the purpose of identification, though it is just barely possible that the difference between "pattern" and some of its previously mentioned relatives may hinge on whether name, clan, or chief is used.

We mention now another example. The index set I consists of the letters of the alphabet together with the diacritical and punctuation marks and (perhaps) the 10 digits. The union of sets

$$\bigcup_{i \in I} X_i$$

is the collection of all marks occurring in some printed or written material. The ability to assign correctly any member of the above set union to the index of its set is a basic part of being literate (though of course, it is not all of it).

Many difficulties arise from our naive and simplistic formulation of the pattern problem, and the principal sources of those difficulties appear to us to be the following:

1. The assumption that the sets X_i are pairwise disjoint.
2. Existential definition of patterns rather than some operationally effective method of assignment of x to its pattern $i = i(x)$.
3. Lack of hierarchical stratification of indexings.
4. Lack of redundancy.
5. Static recognition rather than dynamic training or learning to recognize.

Not just books but veritable libraries have been written on these topics, and it would be impossible to give here anything approaching an adequate discussion of them. Some few hints will be found in the mentioned source [96, pp. 355–370] and many more in the references given there. Having stated our point (which is the connection of sets, indexing, ordinals, and patterns), we close with some scattered remarks on points 1–5, principally with reference to bypasses and, especially, i/c bypasses.

Point 1 arises for various reasons: threshold phenomena, looseness of judgment, errors, inability to recognize sufficiently small differences, and even the Heisenberg uncertainty principle. Generally, it is related to the question of whether bypasses are loss-free or lossy. If this is applied to i/c bypasses, one comes to ask how the rigidly binary disjunctive character of $\in$-relation (either $x \in X$ or $x \notin X$) could be modified by being made more elastic. At a further remove, one might even reach the distinction between qualitative and quantitative differences and the possibility of interpolating in between these two strict extremes.

As to point 2, an effective method will be often based on statistical decision theory. But it has already been suggested earlier that i/c bypasses form a possible conceptual framework for introducing a statistic, such as the average.

Under point 3, it may be noted that the elements of the indexing set I could form a disjoint family of new sets I_j, with a second-order indexing set J. The elements of the latter would then be "patterns of patterns," and the process may be continued. There is an obvious connection with bypass levels, and in particular, with levels between which i/c bypasses of different orders might occur. With reference to patterns, some "high order" patterns might be associated with deep properties, such as style or meaning. Some aspects of hierarchical stratification can already be seen in (an extension of) our Scottish illustration: Beside the extremes of individuals and clans, there exist intermediate groupings such as family, tribe, sept, and subclan.

Redundancy occurs when there is more information than is minimally necessary for strict assignment of an individual x to its pattern $i = i(x)$. A type of redundancy may be introduced by splitting the totality T of individuals into pairwise disjoint sets in several ways

$$T \text{ is } \bigcup_{i \in I} X_i, \text{ it is also } \bigcup_{j \in J} Y_j, \text{ and } \bigcup_{k \in K} Z_k, \ldots$$

For any u in T we have now several patterns: $i(u)$, $j(u)$, $k(u)$, $\ldots$, and if the secondary patterns $j(u)$, $k(u), \ldots$ are known then it may be easier to assign the principal pattern $i(u)$ to u. In terms of our illustration, X_i may refer to belonging to the same clan, Y_j to wearing the same tartan, Z_k to living in the same geographical region, and so on.

Dynamic training or learning referred to in point 5, depends to some extent at least on a combination of previous features. Thus it may depend on more effective methods of assignment of x to $i(x)$, and in particular, on more refined statistical decisions for such assignment. Then, learning to recognize patterns better may come from overcoming the ambiguities present when the sets X_i are (subjectively?) not strictly disjoint. Also, there is a distinct possibility that learning to recognize may depend on having a deeper hierarchy of indexings and on the ability to create ad hoc hierarchical levels. Finally, the ability to proceed under conditions of less and less redundancy may also be involved.

PART THREE

EXAMPLES IN TECHNOLOGY AND THE NATURAL SCIENCES

Chapter Eight

Miscellaneous Simple Technical Examples

EX. 8.1. A TRANSPORT RIDDLE; PHASE BYPASS

The nonmathematical examples of the bypass principle will be started with the following question, something in the nature of a riddle, which is not quite as facetious as it may appear: How do you transport water in a sieve? To score top points the answer should be: freeze it, carry ice in the sieve, then melt it or let it be melted at the other end. An objection might be raised that it is now not water but something else that is carried in the sieve. However, the question was not how to carry water in a sieve. What matters here is the contrast between "carry," which implies transfer in unchanged condition, and "transport" which we interpret as allowing a change of condition so long as that which was given at the start appears at the destination. An objection similar to the preceding one might apply to telephony but only if we claimed that the telephone cable carries the voice in some literal sense rather than transmitting it.

There is an obvious common element to transporting water in a sieve and voice over a cable, and to many other examples of transport bypasses: a suitable transformation of that which is to be transported so as to adapt it to the transportation channel. It is precisely this transformation S at the one end, and its inverse S^{-1} at the other, that make up the outer part $S \ldots S^{-1}$; together with the actual transportation T in between, the full bypass STS^{-1} is made up.

An unacceptable answer to our initial question would be: Grease the sieve, or otherwise close up its pores, and then carry water in it. Even if the sieve were to be de-greased at the other end, for the transportation purpose it would have been replaced by something else, in effect by a vessel. There is an alternative acceptable method which depends on the addition, and then removal, of gelatin or a similarly acting substance, a small quantity of which will bond a large amount of water to a jelly. This alternative method of water transport is the basis of a well-known treatment of constipation by irrigating the large intestine and the colon.

The gel–carry–ungel type of bypass suggests other and more serious therapeutic applications. The idea is to attach a chemical, or perhaps a molecular group, to a suitable carrier, and to produce controlled release at the proper place and time within the ailing organism. Some superficial analogy may be observed here to the masking or stenciling example of chemical bypass mentioned in Chapter 1. This analogy is perhaps reinforced by the example of pill-coating, as used to protect rather than to sweeten. The coating is added, the pill is swallowed and moves through some parts of the digestive system being protected against the chemicals present there, then at the proper place the coating dissolves. The idea of using bypass stacking for more effective protection and better therapeutics seems to be attractive here.

Both methods of solving our water and sieve riddle are bypasses, though different ones: If the freeze method is schematized as $W = STS^{-1}$, then the gel method is $W = S_1 T S_1^{-1}$. This shows that a given W and T may be conjugate under two, or more, quite different operations. In our example W is the (problem of) transporting of water and T stands for being carried in a sieve; S is freezing and S^{-1} melting, S_1 is gelling and S_1^{-1} ungelling. Using transport terms, we can express such multiple conjugacy as follows: The same transportation problem may be solved in different ways even if both the transport channel and the transportation conditions remain the same. In other words, the stuff to be carried can be adapted to the given channel in several different ways.

It is noted that both solutions to our riddle operate on a bypass conversion of liquid—solid—back to liquid. The reverse bypass, solid—liquid—back to solid, has considerable applications. One of them concerns transporting and processing of certain ores and minerals. To expedite things, it may be best to powder the solid substance as close as possible to the place where it is mined and to add water. In effect, the solid ore has been converted into a thick liquid or slurry. This may be pumped to a first-level, or preliminary, processing plant that produces by some thermo-electro-chemical means an enriched solid, for instance, pellets or bars. These are then sent somewhere else for final processing and refinement; one result may well be an increased efficiency in transport.

An important instance of the solid—liquid—solid bypass is biological; it concerns food metabolism. We note that many of the principal transportation networks of the animal body are pipes (often quite narrow ones), that the functions of the body organs are relatively specialized and the organs themselves spatially separated, and that food is often eaten in solid form. Also, the process of eating and digestion is a matter of substance transport followed by, or united with, chemical-energy transport. So, it makes sense that much of biological transport should operate on a solid—liquid—solid bypass, in respect to respiratory and circulatory functions we meet, of course, a gas—liquid—back to gas type of bypass.

Our last specific example is a use of the solid—gas—solid bypass to separate the uranium isotopes. Let the two forms of uranium be U and U'; one begins by forming the gas, uranium hexafluoride, which contains both types of molecules, UF_6 and $U'F_6$. Their separation is based on a slight density difference, and is done by an ultra-centrifuge or by a mass-spectrograph. Finally the desired isotope U' is obtained by converting $U'F_6$ back to solid form.

The generic schema for all our examples may be called a phase bypass: $F_1 \rightarrow F_2 \rightarrow F_1$, where F_1 and F_2 stand for two different phases out of the three standard ones: solid, liquid, gas. The form $F_1 \rightarrow F_2 \rightarrow F_1$ is very inexact; it might be replaced by our usual STS^{-1}, with the following interpretation: S is the phase-change process from F_1 to F_2, T is some process applied then in the phase F_2, and finally S^{-1} is the reconversion to the initial phase F_1. The phases appear now as what we have before called levels. The form STS^{-1} is perhaps better than $F_1 \rightarrow F_2 \rightarrow F_1$, but it still suffers from the shortcoming of not displaying the substances participating at various stages. There is here a certain similarity to what we have called skeletal and full forms of certain i/c bypasses in the last example of the previous chapter: those used in our Scottish illustration. The sort of thing we are distantly and very provisionally approaching in this paragraph is some kind of bypass formalism and symbology, perhaps even some kind of bypass-calculus; these may come, eventually.

In the few examples given in this section, and in many more similar ones, the matters of bypass extension in length and, especially, bypass stacking in depth, may be of interest, sometimes perhaps even of central interest. It is in such matters that our central point, stated in Chapter 3, arises: the use of conjugacy principle to form, first, analogies between different and even widely disparate disciplines, and then to suggest possibilities for borrowing and sharing of techniques.

EX. 8.2. TRANSPORT, COLLECTING-DISTRIBUTING NETS

The subject of this example may appear simple since, after all, it concerns carrying something from A to B. However, even if simple, it is plainly far from trivial from the scientific, technological, industrial, or political point of view. It may be not entirely farfetched to begin by noting certain recent anthropological claims, for instance, those in Reference 75. According to them the shift from eating food at once, on finding or killing it, to carrying it for later consumption and sharing, was important in producing the change from advanced pre-humans to early men. To elaborate this a little further, it is claimed that the above primitive transport of food, together perhaps with the equally primitive transport of minerals suitable for making stone tools, helped to establish such fundamentals of human society as family units and specialization of labor.

Yet some animals also carry food to share it, at least with their mates and young, birds bring materials for their nests from afar, and it has even been observed that chimpanzees will occasionally transport sticks over some distance to dig out termites. One further simple notion is needed to shift a transport situation that could still be stretched to apply to animals, to a distinctively human one: container. It may be speculated that the first containers were loosely gathered hides of eaten animals, large leaves, or hollow gourds; that point is unimportant here. What matters is that once the idea of carrying is joined with the use of a container, we get the germ of two simple and related, but fundamental, bypasses

$$\text{pack—transport—unpack}$$
$$\text{collect—transport—distribute.} \tag{1}$$

Another essential bypass-germ might be noted

$$\text{pack—store—unpack.} \tag{2}$$

But this one belongs to a different class, being a temporal rather than spatial transport.

To pass from the anthropological scale of millions of years to one of thousands, we have a remarkable catalog of human achievement in a long speech from the ancient Greek drama *Prometheus Bound* by Aeschylus. Here Prometheus represents himself not merely as the bringer of fire but as the great culture-hero, who enumerates at length his gifts to men. Together with the use of numbers, letters, metals, medicines, architecture, and prophecy, he mentions two gifts specifically related to transport: the domestication of horses and the invention of ships.

To jump at once from the beginnings of transport to its role today, we need merely observe the huge amounts of goods, articles, objects, and passengers that must be transported these days to support our modern civilization. This alone explains why ministries of transport are parts of governments, departments of transport studies are parts of universities, and transport-law units parts of law faculties. A large part of the problem of transport can usually, though not always, be stated simply: how to carry the biggest possible cargo in the most efficient way. Here, for "most efficient" we can usually, if not quite always, substitute "cheapest." At any rate, transport problems go naturally with various types of optimization.

Let us quote now three of the economic facts of life in transport. First, any change from one mode of transport (truck, airplane, train, ship) to another, or even from one vehicle to another one of the same type, is lengthy, cumbersome, and expensive. So, one avoids transshipping, or trans-loading A phrase from the modern jargon is: transshipping is labor-intensive. The second item goes, to some extent, in the opposite direction: Bulk transport, by ship especially, is one of the cheapest methods. Therefore, it may pay to load up and unload under some conditions. The third fact is very simple: For its efficient use the transporting vehicle should be filled to its capacity. A simple but suggestive mathematical phrase is: The packing fraction should be close to 1.

To optimize under these and some other realistic conditions, the transportation experts came up with what they variously describe as "modular transport," "containerization," or even "the container revolution." The idea is to standardize load by dividing it into equal containers that fit together well. Such containers are filled up locally and then collected and shipped, or transshipped, as wholes. A truck or a plane might carry one, two, or three such containers, a railway flatcar two to four, and a ship several hundred or more. The very considerable savings resulting from such a mode of containerization transport are enthusiastically described in Reference 54.

What is the principle behind such containerization? From the bypass point of view, it is simply described as bypass stacking to the depth of three levels instead of two, with a standardization at the middle level:

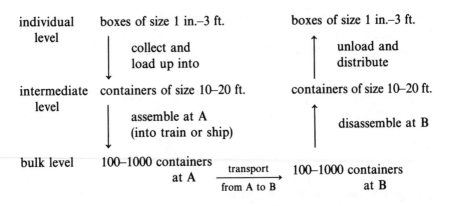

A particularly clear use of such bypass has been sometimes called "piggyback" transport: Here the containers serve as complete vans of trucks, they are taken to a railway station, and the vans are then moved as units, two or three to a flatcar.

The two basic bypasses (1) may help in mathematical modeling of transport problems and they may even suggest mathematical techniques for solving such problems. As an example we consider a simple collecting—transporting—distributing network, illustrated in Figure 1. Two groups of points are given in the plane: $x_1, \ldots, x_n; y_1, \ldots, y_m$, together with nonnegative numbers, or weights, $w_1, \ldots, w_n, W, v_1, \ldots, v_m$. The object is to locate points x and y so as to minimize the function

$$C(x,y) = \sum_{i=1}^{n} w_i |xx_i| + W|xy| + \sum_{j=1}^{m} v_j |yy_j|, \tag{3}$$

where $|ab|$ is the length of the straight segment ab. The idea is that something is produced at the points x_i, and from these production centers it is to be

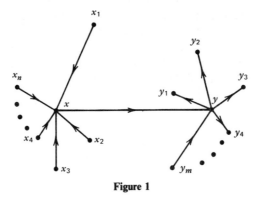

Figure 1

sent by one type of transport to the collection center x. From x the collected total is to be sent by some sort of bulk transport to the point y. Finally, y serves as the distribution center to the consumption centers y_j, and this distribution may be by yet another type of transport. The constants w_i, W, v_j take care of the amounts transported and of unit moving costs, and $C(x,y)$ is then the total cost of the operation.

The problem could be complicated by introducing certain features of more realistic transport. For instance, the centres x and y may not be arbitrary in the plane but may be confined instead to certain regions, curves, or even discrete point-sets. This would certainly be the case if the bulk transport from x to y is by existing railway lines or water transport facilities. Of course, the volume of the operation might be big enough to justify, economically at least, building new rail or canal links. Further, one might modify the assumptions leading to (3) so as to introduce "containerization" by adding suitable trans-shipping costs.

Finally, there is a deeper modification in which the simple "one-stage" network of Figure 1 is replaced by a hierarchical "two-stage" network. The idea is that instead of n production centers and m consumption centers to be optimally connected, there are now n clumps of production centers and m clumps of consumption centers. The optimal connection is to be done in two stages: There is one secondary collection point for each production clump, and similarly one secondary distribution point for each consumption clump. All the secondary collection points are to be joined to one primary collection point, from which there is some bulk transport to the single primary distribution point. The same two-stage process is repeated (but in reverse: as in bypass!) at the distribution end. Just as Figure 1 shows a simple bypass idea, by the very direction of the arrows if by nothing else, so now the two-stage operation is schematized by a stacked bypass with three levels:

individual level	production centers $x_{i1}, \ldots, x_{if(i)}$	consumption centers $y_{j1}, \ldots, y_{jg(j)}$
	collect to $\downarrow$	$\uparrow$ distribute to
secondary level	secondary collection centers $x_1, \ldots, x_n$	secondary distribution centers $y_1, \ldots, y_m$
	collect to $\downarrow$	$\uparrow$ distribute to
primary level	main collection center x	$\xrightarrow[\text{transport to}]{\text{bulk}}$ main distribution center y

Next, suitable weights and linear cost functions are assigned to each of the five links of the preceding bypass diagram. We then get a total cost function, as in (3), but with five terms (for the five links of the preceding stacked bypass) instead of the three terms of (3) for the three links of the simple bypass of Figure 1:

$$C(x_1, \ldots, x_n; x; y; y_1, \ldots, y_m) = \sum_{i=1}^{n} \sum_{p=1}^{f(i)} w_{ip} |x_{ip}x_i| + \sum_{i=1}^{n} W_i |x_i x|$$

$$\tag{4}$$

$$+ W|xy| + \sum_{j=1}^{m} V_j |yy_j| + \sum_{j=1}^{m} \sum_{q=1}^{g(j)} v_{jq} |y_{jq}y_j|.$$

As the preceding discussion suggests, bypass ideas might be of some use in transportation-net problems in particular and in operations research in general. We shall confine ourselves to some remarks on the problem of minimizing $C(x,y)$ in (3), which is complicated enough in spite of its apparent simplicity. The complications arise mainly from the possibility of boundary minima [one type: x coincides with y; another type: x or y (or both) may occur at a point x_i or y_j]. Of special interest might be the critical value of the bulk-transport cost W, above which the first type of boundary minimum might occur or, in simple terms, the cost above which bulk transport prices itself out of existence.

It might be observed that the problem of minimizing (3) contains as a very special case the so-called Weber problem [83, 139] of spatial economics, on the optimal location of industries. To get the Weber problem out of ours one simply takes $W = v_1 = \ldots = v_m = 0$ in (3).

Next, we outline briefly an approach to minimizing (3) based on the elementary properties of certain curves called polyellipses [97] that generalize the ordinary ellipses. A polyellipse is the plane locus of points p for which the weighted sum of distances to n given points $x_1, \ldots, x_n$ (the n foci) in the plane is constant:

$$\sum_{1}^{n} w_i |px_i| = c.$$

The constant c must satisfy $c \geqslant c_0$, where c_0 is the minimum of the left-hand side over all p. Aside from an easily characterized exceptional case, the polyellipse $P(c)$ reduces for $c = c_0$ to a point; for $c > c_0$, $P(c)$ is a strictly convex curve, and $P(c_2)$ contains $P(c_1)$ in its interior if $c_2 > c_1$. There is exactly one polyellipse $P(c)$ through every point in the plane.

So, let the optimal collection point x lie on the polyellipse $P(c_1)$ with foci $x_1, \ldots, x_n$; let also the optimal distribution point y lie on the polyellipse

$Q(d_1)$ with the foci $y_1, \ldots, y_m$. Then only the simple middle term $W|xy|$ in (3) remains to be minimized, hence xy is the shortest straight segment joining the two polyellipses, and, as such, it is normal to each polyellipse at the common point. As is shown in Reference 97, there are good algorithms for drawing families of confocal polyellipses. Thus the problem is reduced from the four unknowns (i.e., the coordinates of x and y) to the two unknowns c_1 and d_1 and now a reasonable minimum-hunting algorithm is not hard to set up.

One feature of the preceding transport-net minimization problems is sufficiently important to merit special further discussion and even a separate name. It depends on the "trunking" process or, equivalently, on the collection and distribution; we propose to call it the SL-bypass (here SL is a mnemonic for square-linear), alternative possible names would be MA (multiplication-addition) or SOM (saving an order of magnitude).

Like some other bypasses, past or to come, the SL-bypass is well masked by being completely out in the open. Consider our optimization problem of minimizing the cost function (3), which relates to Figure 1. Suppose that it had not occurred to us to use the bypass:

collect (to x)—transport (from x to y)—distribute (from y).

Assume also that what is made at each production center x_i is in demand at every consumption center y_j. It is then necessary to transport from each point x_i to each point y_j; thus, instead of the $m + n + 1$ segments of the network of Figure 1, there are mn separate segments. So, if m and n are at all large, we have many more segments and a much longer and more complex total net. One could even take this as yet another justification of our claim that the conjugacy principle is a complexity-reducing device. To continue with our example, it might happen that the production and consumption centers coincide: They are just N given points. Then the separate interconnection scheme requires

$$\binom{N}{2} = \tfrac{1}{2}N^2 - \tfrac{1}{2}N$$

segments, while the collecting and distributing net calls for at most

$$2N + 1$$

of them. The comparison of those two magnitudes justifies the name SL. Of course, it might happen that trunking *all* n production centers to a single collection center is inconvenient: The number n may be too large for proper handling and service; the same may apply at the distribution end. This is certainly the case with very large communication nets such as the telephone, or the interurban and suburban railways. It may then be preferable to use the stacked SL-bypass of a two-stage (or even multistage) collecting and distributing net leading to minimizing the cost function C in equation (4).

The occurrence of the SL-bypass is more widespread than might be at first suspected. Thus, for example, we have alluded in the fifth preliminary example of Chapter 1 to the barter and the money economies. It was suggested that the use of money (for exchange) rests on a bypass: sell one's goods or services for money—keep the money—buy with money the goods or services required. Let there be N commodites, then the complete barter system would require again $N^2/2 - N/2$ exchange ratios, whereas the money system, viewed as an SL-bypass, requires only N prices. Some considerations regarding interest, inflation, and roles of money other than as an exchange medium might even distantly hint at something like our collecting—transporting—distributing bypass, simple or stacked.

It has been sometimes observed that there is a certain functional similarity between money and language. We might even expand it to claim a likeness of money, language, and transport. Such likeness may come from their all being bypasses. The money/language matters are mentioned by Leibniz, and a whole article on this appears in a recent Leibniz memorial volume [36]. However, one of the best statements of the money-language analogy antedates Leibniz; it is due to the seventeenth-century English savant and divine John Wilkins, In Chapter 1 of Reference 141, he says, concerning communication by speech, writing, or gesture: "And so this subject belongs to the mint of knowledge; expressions being current for conceits, as money is for valuation."

Such analogies make it tempting to speculate on bypasses in language, and with reference to our present topic, on possible SL-bypasses in language. The possibility alone is illustrated by remembering the huge multiplicity of tongues and dialects. Some 2000–5000 are estimated to have existed, and the number in current use is still quite large. Hence the number of separate dictionaries needed for complete translating is very large; in particular, it is a *quadratic* function of the number of languages. However, if some "comprehending," or "union," or "pre-Babel" language existed in addition to the natural ones, then the necessary number of dictionaries would be only a *linear* function of the number of languages.

We finish this example with an imaginative bypass of the telephony type, for the purposes of imaginary transport, used as a stage prop by certain conscientious science-fiction writers. It is usually called by some such name as "transmat" (i.e., transmission of matter). To move something over inter-planetary or interstellar distances, the object is scanned atom-by-atom, the scan gets coded into light waves or other form of radiation, beamed to its destination, and there reassembled, again atom-by-atom. Some modern developments in lasers and holography make the transmat slightly less improbable; a likely candidate for this type of transport would be transfer of single interesting molecules, say from a crewless space ship back to Earth. It is of interest to note that the commonest ploy of science fiction to beat the relativity barrier and so produce an FTL (faster than light) drive is also a pure, if entirely imaginary, bypass: enter from "ordinary" space into "hyperspace," move in the hyperspace at the velocity of suitably many light-years per hour, then reenter the "ordinary" space.

EX. 8.3.   MAPS

This example concerns maps and their use so that it still has a connection with transport. However, the important analogy now is with Ex. 4.10, concerning manifolds. It may be something of an accident that the words "map" and "mapping" and then even "chart" and "atlas" have been borrowed as technical terms into modern mathematics. Yet there may be here a certain instinctive soundness in the choosing of names; at any rate, we shall claim that the analogy is more than superficial.

In the previous example something definite and material was moved from A to B, and this led to one kind of bypasses. Now, we are all well acquainted with the concept and use of maps; what is the bypass here, if any? It is precisely at this point that the analogy to manifolds begins to apply. A simple schema of map-bypass is just this:

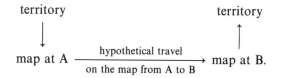

Thus it might be said that it is structure which is being transported here, just as with manifolds. For slightly greater detail, we might observe that "manifold" corresponds to "territory." and since a manifold is, to some extent, a strange country, we use a local map; this map represents the strange country on something familiar, that is, a Euclidean space. Bypass stacking occurs here, and to about the same depth of two to four levels as in the previous example. The stacking of map-bypasses is best illustrated on the scale of maps.

It is the structure that is transported now, and sometimes there may be too much of it. For instance, if we are in the city A and intend to drive to the city B, we might start by using a detailed map of A and its immediate surroundings; in manifold terms, it is the local map of the territory near A. We use it to take us "out of A" and to put us on the right road. Thereafter we need much less structure, essentially only an indication of the road we want. So, a much cruder large-scale road map will do to direct us, mainly to take correct turns at principal intersections. When we approach B, we may switch over again to a more detailed map: the local map of the city B. Or, if driving from A to B we get lost, we might use some local map to "place ourselves" by means of local clues, and then return to the large-scale road map. Such stacked bypasses are of the type:

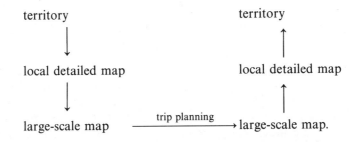

Bypass extensions in length, according to the schema (C₄) from Chapter 1, occur here too: for instance, when we stop for a "mid-course correction." We observe then where we are in the land, transfer ourselves on to the map, check the true course, transfer ourselves back to the territory by checking some landmarks of the true course, and we continue. Finally, bypass inversion is also present, in correcting and updating of maps: We start then from some old map, pass over to the actual territory, observe the corrections or addenda, and then go back to the map to place the corrections on it.

This is a good place to make some preliminary observations on the use of metaphors, similes, and analogies. It is not entirely unprofitable to recall that "analogia" means "proportion" in Greek (as was already noted in Ex. 4.0) and that "metaphor" means "transfer." Such matters belong to language and its uses and will be discussed at some length later. But we observe now, in connection with maps, such figures of speech as "terra incognita" for something unknown (to us), "putting oneself on the map" for becoming known or notorious, "being on the right road" for taking the correct course to achieve something or to get somewhere, and so on. A certain stacking of bypasses seems to be involved here. We start with a given general situation that we wish to illuminate; it may remind us, by some association, of something else, in this case of the map-bypass. A figure of speech is used, for instance, one of the three quoted. It may be useful to analyze just how the point is made, and a possible analysis could run like this: transfer first from the given context or situation to the initial term "territory" of the map-bypass, invoke implicitly that bypass by the very making of that figure of speech, then transfer from the "territory" as the last term of the map-bypass back to the given situation. If such an analysis seems primitive, it might be granted to be a step in the right direction, inasmuch as metaphoric use of language is complex while bypass is claimed to be a way of resolving complexity.

Maps and the associated bypasses may be of some interest in psychology. For, if we allow the existence of something like "mental maps," then these might play a considerable role in behavior, especially in the planning of future behavior. An interesting reference here is to the title of a text in psychology [100]: *Plans and the Structure of Behavior*, and to the obvious similarity of "plan" and "map." It might even be claimed that a good brief description of intelligence is that it serves to produce correct behavior in a strange territory. Such ability seems to call for a facility to produce fast "mental maps," temporary ones perhaps, which are sufficiently accurate as general guides. It may be that such "fast mental maps" are produced just as the ordinary maps are: in patches and by fitting patch-maps together on their overlap. If this were granted, then the analogy with structure-transporting bypasses of Ex. 4.10, on manifolds would be greatly reinforced. Special attention is drawn to the possibility that such fast *ad hoc* maps may get transferred, on proving themselves correct, to some semipermanent or permanent mental map or atlas.

There is a fairly obvious connection between our "mental maps" and the subject of hierarchical patterns and hierarchical pattern indexing from the end of the last chapter. Presumably, a deeper hierarchy is involved in more accurate, and in faster, production of temporary maps. This is so because a deeper and finer set of categories for discrimination of detail is then available.

The idea of using maps in psychology is by no means new. There is an old psychological theory due to K. Lewin [86, 87] in which the central terms are differentiation and dedifferentiation. The first process goes with growing up and maturing and it means, roughly, refining one's mental map, or one's picture of the world, so as to give truer image and better detail (or finer structure). The second process is a reverse pathological condition that Lewin associates with mental difficulties and conditions. There is also another attempt to emphasize psychopathology as a failure associated with human mapmaking and human map using. It comes from a strange discipline christened by its founder A. Korzybski "general semantics" [82]. Korzybski has "territories" and "maps" and even higher-order "maps" ("maps" of "maps") and he formulates a general motto: "The map is not the territory." With this, he attempts to present certain types of insanity as a confusion of a (private) map and the (public) territory.

In concluding this example we mention briefly a concept that appears to be related to map: blueprint. It is not clear whether blueprints involve bypasses, even if it is granted that maps do. One might say that blueprints serve to bridge the gap between the conceptual reality in the designer's mind and the physical reality at the producer's hands, but this is a bit farfetched. However, one aspect of blueprints is analogous to what we have called the Apollonian bypass in Ex. 4.9. This concerned the ancient Greek synthetic geometry of conic sections and was schematized as $W = STS^{-1}$. Here W denotes doing something in the plane (which Apollonius in his time could *not* do), S is the passage from plane to space, T stands for working in space, and S^{-1} is the return to the plane. In the terms explained there, the Apollonian bypass is coded as 2/3/2. Now, one of the purposes of blueprint use is to show in *two* dimensions what happens in *three*. Hence a certain 3/2/3 bypass appears to be associated with blueprints.

EX. 8.4. TEMPORARY HOLDING DEVICES

These were sketchily introduced in the fifth preliminary example of Chapter 1 as scaffoldings, splints, vises, clamps, jacks, levers, and so on. Their object is temporary positioning, immobilizing, lifting, holding, stiffening, detaining, and so on. The basic bypass may be summarized here as clamp on—hold—unclamp, put on—hold—take off, hold—keep—let go, apply—keep on—remove, and a variety of other similar three-verb strings. In the preliminary example of Chapter 1, we considered the matter of scaffoldings, in the form

$$S_1 S_2 \ldots S_n T_1 T_2 \ldots T_n S_n^{-1} \ldots S_2^{-1} S_1^{-1}, \tag{1}$$

where S_i stood for the ith floor scaffolding, or perhaps for the operation of putting S_i on top of S_{i-1} (S_0 being the ground level), and T_i meant building the ith floor of an n-floor tower (on top of T_{i-1}). If only the scaffoldings S_k for $k \leqslant i$ are needed to build T_i, then in the bypass (1) there is a restricted commutativity

$$S_i T_j = T_j S_i \qquad \text{if } i > j. \tag{2}$$

So, it might be asked: In how many ways can (1) be rewritten, subject to the conditions (2)? By the standard techniques of reflection and the balloting theorem [49, Ch.3] it is found that the number of different ways of building an n-story tower, with respect to sequencing the scaffold erection and successive floor building, is then

$$\frac{1}{n+1}\binom{2n}{n}.$$

Other constraints besides (2) could be imposed, for instance, that the total scaffolding height never exceed the already built part of the tower by more than q floors. This leads to a variety of combinatorial problems concerned with the number of ways of rewriting the string (1). Some such problems might even acquire a certain practical interest if probabilities could be assigned to various safety features or to the chances that either the scaffolding parts or the building materials might not arrive on time. It might then be desirable to maximize safety or to minimize the expected time lost in waiting. Other, though remoter, possibilities of similar type might arise from some items in the theory of scheduling of unit industrial operations of the bypass type, when some operations have to precede certain others.

For the next instance, the temporary holding device used is the simple lever, for example, a crowbar. Thus the basic bypass here has the outer part $A \ldots A^{-1}$, where A stands for applying the crowbar, to raise a load for instance, and A^{-1} stands for removing the crowbar. We shall consider in some detail a moderately elaborate problem, not so much for its own sake but to illustrate two points. The first one is to show bypass stacking in the form of literal and physical stacking. The second point concerns our earlier contention that the conjugacy principle is something of a feature of natural software. In this connection our object will be to show how to combine simple bypasses so as to "program" the job to be done.

A heavy weight, say a lead-lined, armored safe weighing several thousand pounds, is originally in the position M shown in Figure 1a, at the foot of a ramp R. The object is to raise it to the position M_1 in which a heavy load, say uranium, is placed inside the safe, and then to lower the safe back to its initial position M (say, for return transport). The operation U of placing

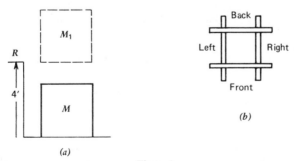

Figure 1

the load inside the safe is treated as a single unit that does not concern us further. The raising and the lowering are to be done by primitive means: three men, two strong crowbars, and sufficiently many thick wooden boards, like railway ties. For definiteness, each tie is 3 in. thick and of approximately the same length as the base of the safe (which is, say, a flat square).

The simple initial schema for doing the job is already a bypass, naturally RUR^{-1}, where R stands for raising from M to M_1 and R^{-1} for the reverse lowering. This already cuts down the amount of "planning": Only R has to be produced out of simpler bypasses; R^{-1} is then taken care of automatically. The operation R is synthesized by stacking the ties under the safe in alternate directions, as is shown in Figure 1b. The full basic bypass is now amended to AIA^{-1}, and where A stands for (synchronized) applying of two crowbars under suitable two corners, and I for inserting a tie under the safe. If the diagram of Figure 1b represents the lowest layer of the two bottommost pairs of ties, then the operation of laying the four of them may be schematized in the obvious terminology as

$$(AIA^{-1})_{l_1}(AIA^{-1})_{r_1}(AIA^{-1})_{f_1}(AIA^{-1})_{b_1}.$$

To produce full R, the preceding operation is repeated eight times, and we get

$$R = \prod_{i=1}^{8} (AIA^{-1})_{l_i}(AIA^{-1})_{r_i}(AIA^{-1})_{f_i}(AIA^{-1})_{b_i} \qquad (3)$$

each successive pair of ties being laid on top of the previous one.

Two minor remarks need to be made here. First, there is a certain limited freedom of sequencing the placing of ties; this reflects in a certain limited type of commutativity in the product (3) [just as we had the limited commutativity (2) with scaffoldings]. However, this seems to be of no interest here. Second, the new layers of ties could be inserted below the previous ones, so that both the safe and the partial scaffolding under it are rising together.

To produce the inverse, R^{-1}, we observe the usual rule that the inverse of a product is the reversed product of the inverses, and we get from (3)

$$R^{-1} = \prod_{i=8}^{1} (AI^{-1}A^{-1})_{b_i}(AI^{-1}A^{-1})_{f_i}(AI^{-1}A^{-1})_{r_i}(AI^{-1}A^{-1})_{l_i} \qquad (4)$$

since

$$(AIA^{-1})^{-1} = AI^{-1}A^{-1}.$$

Here, of course, I^{-1} denotes removing the tie from under the safe. With the initial schema RUR^{-1}, together with (3) and (4), the whole job has been broken up into its simple constituent bypass parts.

The very large membership of the class of temporary holding devices would enable anyone to quote countless further instances of more or less interest. We shall end up with two rather unexpected examples, both of which qualify quite well, as it seems to us, as temporary holding devices. The first one concerns a certain aid in the problem of composition. Although certain writers are able to produce a finished copy at once, many of us have to rewrite, and re-rewrite, our work in a series of successive approximations. In producing such improved later versions out of the earlier ones, the temporary holding structure technique may be found psychologically useful, somewhat on the analogy to scaffoldings. That is, the work may be facilitated by adding to the previous draft certain paragraphs, sketches, or even marginal remarks or doodles, which act as a temporary glue. Once they have performed their job of helping to induce better coherence, and the next approximation is produced, they are removed.

The last temporary holding device we discuss may surprise: the jail. Yet, it is a temporary holding device in a very literal sense. Since there is such a thing as jail-within-the-jail, solitary confinement, it may even be claimed that there is a sort of bypass stacking here. Bypass extension in length is rather obvious. Some interest might attach to a certain inversion of the jail-bypass. It seems that the one-time, or occasional, offender often regards his time according to the direct bypass

commitment to prison—incarceration—release

while the habitual criminal frequently marks his time by the reverse bypass

release ($\#i$)—freedom—commitment ($\#i + 1$) to prison.

That is, in both cases there is a round trip between the "levels" of freedom and prison. In the first case the interval of imprisonment is a passage between two free periods. In the second case the period of freedom, used as the time to plan and pull off another job, is a passage between two sojourns in jail.

The above phenomenon may be called a bypass switchover: The same two levels are involved, but the direction of the bypass round trip between them reverses. We shall suggest that something like a bypass switchover appears in the basic human act of sleep or more precisely, in the development of sleep from infancy to maturity. Now, it is not clear whether the sequence of falling asleep—sleeping—waking is a bypass or not, but some bypass features at least seem to be present. It is tempting to speculate on the presence of some further bypass features in sleep: sleep stages (or levels) intermediate between full wakefulness and deep sleep, bypass stacking, presence or absence of dreams at various stages, connections with anesthesia, and so on.

An important general observation must be made here. This concerns many of our examples, especially those outside mathematics, and most particularly those outside science and technology. In such "distant" examples the label of bypass placed on them does not mean that the thing in question is a bypass, wholly a bypass, and nothing but a bypass. The claim of bypass is emphatically not meant to constitute the whole truth of it. We mean only that some aspect, which might well turn out to be incidental (though it need not be), may be profitably modeled in terms of bypass.

This is the case with the development of the function of sleep in humans from birth to maturity; we suggest that it may be regarded as a bypass switchover. The newborn sleeps most of the time, waking now and again for food; later on sleep becomes polyphasic (i.e., many sleep periods per 24 hours), then diphasic, and eventually monophasic. The infant's existence is sleep interrupted by some wakefulness; the adult state is wakefulness with one period of sleep per day. A partial reverse or the switchover may occur in some pathological conditions or, to some extent, in senescence.

Our last remark returns to the topic of jail as a temporary holding device: prisons, prisoner-of-war camps, and concentration camps seem to provide an example of yet another bypass. There are the obvious two levels: life "in" and life "out." To make the "in" bearable, or in some extreme cases to increase survival probability and even to make the daily routine of "in" possible, the inmates have to give up certain esthetic, ethical, and moral characteristics of the normal or "out" existence. It is interesting to hypothesize about the sequence in which those characteristics are lost, and the sequence in which they are regained (after release).

Chapter Nine

Examples Relative to Physics and Technology

EX. 9.1. THE HAMILTON–JACOBI THEORY OF MOTION

In this theory [58], a mechanical system is described by its position variables $q_1, \ldots, q_n$ and the associated momentum variables $p_1, \ldots, p_n$, all of which depend on time t alone. Differentiation with respect to t is written in the Newton form as

$$\dot{p}_i = dp_i/dt, \qquad \dot{q}_i = dq_i/dt.$$

There is the Hamiltonian function

$$H = H(q_1, \ldots, q_n; p_1, \ldots, p_n; t)$$

representing under certain restrictions [58, p. 211] the total energy of the system. The basic equations of motion, known as Hamilton equations, are derived from a variational principle; they are

$$\dot{q}_i = \frac{\partial H}{\partial p_i}, \qquad \dot{p}_i = -\frac{\partial H}{\partial q_i}, \qquad i = 1, \ldots, n. \tag{1}$$

The Hamilton–Jacobi (HJ) procedure is described concisely from a physicist's point of view in [58], at length in Jacobi's original memoir [76], and in a modern differential-geometric setting in [125]. It reduces solving the system

(1) of $2n$ ordinary first-order differential equations to solving one first-order partial differential equation. It is shown in [58] on physical examples how this reduction may lead to the solution of the original physical problem.

The idea behind the HJ procedure is a bypass based on the canonical transformation

$$Q_1 = Q_i(q_1, \ldots, q_n; p_1, \ldots, p_n; t),$$
$$P_i = P_i(q_1, \ldots, q_n; p_1, \ldots, p_n; t) \qquad i = 1, \ldots, n, \tag{2}$$

to new variables Q_i and P_i. The transformation (2) is canonical if there is a new Hamiltonian

$$K = K(Q_1, \ldots, Q_n; P_1, \ldots, P_n; t)$$

such that the transformed equations of motion have the same form as (1):

$$\dot{Q}_i = \frac{\partial K}{\partial P_i}, \qquad \dot{P}_i = -\frac{\partial K}{\partial Q_i}, \qquad i = 1, \ldots, n. \tag{3}$$

The bypass itself is schematized as follows:

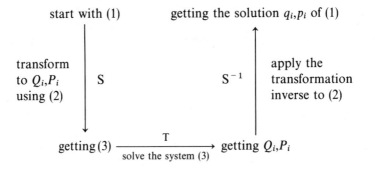

start with (1)   getting the solution q_i, p_i of (1)

transform to Q_i, P_i using (2) — S S^{-1} — apply the transformation inverse to (2)

getting (3) $\xrightarrow[\text{solve the system (3)}]{\text{T}}$ getting Q_i, P_i

We have here, in addition to Ex. 5.10 of Chapter 5, another strong confirmation of Jacobi's claim of the power of the inversion technique [24, p. 557]. Unlike with integral transforms where the first part of the bypass is easy and the third one, the inversion, is hard, here it is the first part that is difficult. The reason for this is in the nature of the HJ method itself. It eliminates the middle part T of the bypass by trivializing it—a condition is imposed that the new Hamiltonian K must vanish. Since $K = 0$, equations (3) are trivially solved: Q_i and P_i are constant.

Actually, due to dissymetry with which the positions and the momenta enter, the procedure is arranged somewhat differently. Hamilton's principal function S is introduced as

$$S = S(q_1, \ldots, q_n; P_1, \ldots, P_n; t)$$

relating the old and the new Hamiltonians by

$$K = H + \frac{\partial S}{\partial t}.$$

To make $K = 0$, S must satisfy the Hamilton–Jacobi equation

$$H(q_1, \ldots, q_n; \frac{\partial S}{\partial q_1}, \ldots, \frac{\partial S}{\partial q_n}; t) + \frac{\partial S}{\partial t} = 0.$$

This is the partial differential equation referred to earlier. Once S has been found, the whole problem is solved. The new momenta P_i are constants α_i, and the new positions are constants

$$Q_i = \beta_i = \frac{\partial S}{\partial \alpha_i}, \quad i = 1, \ldots, n, \quad S = S(q_1, \ldots, q_n; \alpha_1, \ldots, \alpha_n; t); \quad (4)$$

the values α_i, β_i are obtained from the known initial conditions that go with the original system (1). The old momenta are given by

$$p_i = \frac{\partial S}{\partial q_i}, \quad i = 1, \ldots, n$$

where S is as in (4), and finally the system (4) is inverted to yield the old positions q_i. The inversion is guaranteed by the nonvanishing of the suitable Jacobian.

EX. 9.2   ADIABATIC, PARAMAGNETIC, AND NUCLEAR-MAGNETIC COOLING. BYPASS CYCLES

Paramagnetic and nuclear-magnetic cooling are physical effects that operate on quantum-theoretic grounds and are used to produce very low temperatures [50, Vol. 3, pp. 35–5, 35–6]. We recall first the property of paramagnetism present in certain salts, especially those of the rare-earth element praseo-dymium. Here the individual atomic magnets line up with an externally imposed magnetic field rather than against it, as is the case in the more common diamagnetism. In adiabatic paramagnetic cooling, a sample of the paramagnetic substance is first cooled by a liquid helium bath to the temperature of the order of 1 K in an externally imposed magnetic field. Then the helium bath is removed and the sample is thermally insulated. Finally the magnetic field is removed, with a suitable rate of decrease. For quantum-mechanical reasons the individual atomic magnets undergo transitions at the expense of thermal energy. The result is a cooling of the sample down to the temperature of the order of 10^{-3} K. The bypass here has the simple form: field imposition—thermal insulation—field removal.

The above method can be combined with a similar bypass procedure to achieve further cooling down to the temperature of the order of 10^{-6} K. For this purpose the cooled-down paramagnetic sample is used for second-stage preliminary cooling, to its own temperature, of another material with strong nuclear magnetism. Then the same technique is used as in the preceding adiabatic paramagnetic cooling, this time operating on the individual nuclear magnets rather than on the individual atomic ones.

The initial stage of the cooling process also arises from a bypass, the one that produces liquid helium. However, here the bypass sequence of magnetic field imposition—thermal insulation—magnetic field removal is replaced by

compression—thermal insulation—(adiabatic) expansion.

The helium gas is compressed and precooled, say with liquid hydrogen, then it is thermally insulated, and allowed to expand adiabatically. In the process the helium gas cools down, and then the compression—insulation—expansion cycle is repeated. Eventually the helium liquefies. This was the method used in 1908 by H. K. Onnes to liquefy helium for the first time [45; see liquid gases].

The occurrence of the word "cycle" earlier is incidental but it suggests a brief look at the possibility of generating cycles out of bypasses. Let us start with the first two symbols ST of a bypass schema STS^{-1}, and let us continue the string ST by adding new symbols at the end so that every three consecutive symbols form a bypass schema. We get successively

$$ST, \ STS^{-1}, \ STS^{-1}T^{-1}, \ STS^{-1}T^{-1}S, \ STS^{-1}T^{-1}ST$$

and here we stop: Since the last two symbols are ST again, a cycle has been generated. This four-symbol cycle is

$$T^{-1} \quad \begin{array}{c} S \\ \circlearrowleft \\ S^{-1} \end{array} \quad T.$$

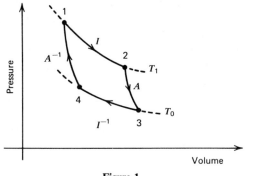

Figure 1

If this is written lineally rather than cyclically, we get $STS^{-1}T^{-1}$; it is curious to note that in group-theoretic terminology this is the commutator $[ST]$ of the elements S and T of a group [84].

The well-known Carnot cycle in elementary thermodynamics might be considered an illustration of such a four-member bypass cycle. We take a quantity of gas at the temperature T_1, and with pressure and volume coordinates at the point 1 in Figure 1. This is allowed to expand along the isothermal I at the temperature T_1 to the point 2, thereby absorbing heat. Then the gas expands from 2 to 3 but adiabatically, along A, till the temperature T_0 (less than T_1) is reached. This is followed by the isothermal compression I^{-1} from 3 to 4, during which heat is given off, and finally the adiabatic compression A^{-1} closes up the cycle to $IAI^{-1}A^{-1}$.

EX. 9.3.   POWER GENERATION, DISTRIBUTION, AND STORAGE

The very idea of a power-distributing network involves a basic bypass. The power is available at one place, in the form of potential energy of dammed-up water, thermal energy of coal, or nuclear energy of uranium, but it is needed at another place or at many other places. Hence it is necessary to find a suitable transmission channel, to transform the power into the form adapted to that channel, transport it through the channel, and retransform it at the other end. The channel is, of course, electricity and the transforming is done by an electric generator. So, we find as a part of the distribution chain the bypass

... S (electric generator) T (transmission line) S^{-1} (electric motor) ... (1)

For instance, one of the first public distribution networks of this type was the Holborn Viaduct power station in London [140], which was opened in January 1882. A steam engine drove an electric generator which produced 60 kW at 110 V D.C., delivered by cables straight to the consumers. A similar system was set up later in the same year in New York on the initiative of Thomas A. Edison [115]; it had six generators with total output of 300 kW, mainly for electrical lighting purposes. These early power plants supplied D.C. power for purely local consumption. The A.C. transformers were developed in 1880–1882 by Gaulard and Gibbs and came into use in 1885–1886 under the the entrepreneurial push of G. Westinghouse; in 1886, A.C. electricity was commercially generated and high-voltage transmission lines were started [28]. An A.C. transmission line at 15 kV, of 175 km length, was in existence by 1891 [115].

The higher the voltage used, the less the power loss in transmitting a given amount of power through a fixed cable. Thus, with the increasing demand for power and the growing length of cables, it became clear very early that high-voltage lines should be used for transmission over any considerable length. Nowadays the figures reach the levels of 300,000–600,000 V. Electric generators cannot produce power at such high voltages, which has two important consequences: the need for electric transformers and the need to change from D.C. to A.C. As it happens, the transformers provide another instance of bypass, according to what we have called bypass extension in depth or bypass stacking: (1) goes over into

$$SUTU^{-1}S^{-1} \tag{2}$$

where S, T, and S^{-1} are as in (1), U is the step-up transformer and U^{-1} the step-down transformer.

However, it has been found that high-voltage D.C. transmission lines have considerable economic advantages over the A.C. ones. Also, there are today efficient, though expensive, A.C. to D.C. and D.C. to A.C. converters capable of handling large powers. Accordingly, the bypass sequence (2) is replaced by

$$SUCTC^{-1}U^{-1}S^{-1}, \tag{3}$$

where S, U, T, U^{-1}, S^{-1} are as before, C is the A.C. to D.C. converter (i.e., rectifier), and C^{-1} is the inverse D.C. to A.C. converter (i.e., inverter). This assumes that the initial electric generator S and the final electric motor S^{-1} both operate on A.C. Otherwise, we have

$$SUCTC^{-1}U^{-1}CS^{-1} \qquad \text{A.C. generator, D.C. motor,} \tag{4}$$

$$SC^{-1}UCTC^{-1}U^{-1}S^{-1} \qquad \text{D.C. generator, A.C. motor,} \qquad (5)$$

$$SC^{-1}UCTC^{-1}U^{-1}CS^{-1} \qquad \text{D.C. generator, D.C. motor.} \qquad (6)$$

We note that in (1)–(6) two chains are not bypasses, (4) and (5), because of A.C.–D.C. dissymmetry.

The preceding examples show an occurrence of bypass for transmission purposes and the complexification of that bypass in passage from direct distribution of generated power to high-voltage A.C. transmission with the use of transformers and then to high-voltage D.C. transmission with both transformers and converters. It will be seen that, in a manner of speaking, each bypass in (1), (2), (3), (6) occurs in series, in a straight-line chain.

The next example of bypass in connection with power is quite different. Here the bypass occurs in parallel to the principal chain, and it is used for the purpose of regulation rather than transmission. The idea of it is simple though somewhat grotesque sounding at first. Since the demand for electric power varies considerably in time, there are slack periods during which the load is low. One control possibility would be to turn down the rate of power generation in the main plant. However, certain power-plants operate more efficiently on an even load, and the following arrangement, outlined schematically in Figure 1, becomes economical. During the slack periods some of the power is used to pump water from a lower reservoir to an upper one, which is the first part S of this bypass. The water accumulates and stays in the upper reservoir, storing up potential energy; this is the middle part T. Finally, at times of high demand, the accumulated water flows back to the lower reservoir through the auxiliary hydroelectric power plant, completing the bypass with S^{-1}.

Our last instance in the context of electric power is another use of bypass

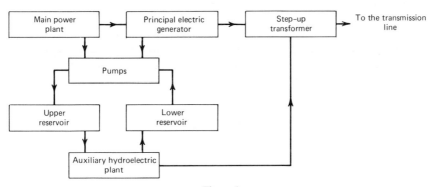

Figure 1

for power storage, as in electric batteries or accumulators. Of the two distinct genera of batteries, the throwaway ones and the rechargeable ones, only the latter is considered. This works on the simple bypass schema

$$\text{charge—store—discharge.} \tag{7}$$

For example, one particular species is the nickel-iron battery, also known as the alkaline battery; it operates on the reversible electrochemical reaction

$$2NiOOH.H_2O + Fe \underset{\text{charge}}{\overset{\text{discharge}}{\rightleftarrows}} 2Ni(OH)_2 + Fe(OH)_2.$$

As in many other bypass examples, we run into various problems of generalized friction and noise, and wear and tear. With the batteries the efficient use is limited by the gradual dispersal and degeneration of the cycled chemicals, by the erosion of battery parts, and by the slow leakage, which eventually discharges the battery even under no external load.

EX. 9.4.   INFORMATION STORAGE, EARLY COMPUTING, MAGNETIC CORE MEMORIES, AND B-H HYSTERESIS

The preceding example concerned the electric battery. Although this could be charged to less than full capacity, for certain purposes it may be useful to regard the battery as a two-state device operating on the bypass: S (charge to full up)—T (store)—S^{-1} (discharge to empty). The middle part T has a passive aspect to it: Nothing particular is done or happens but the feature of usefulness is storing up something, here the electrical energy, so that it will be available later.

Such passivity may be surprising since bypass is often associated with transport, and we think of transport as a motion. Although transport involves both spatial and temporal terms, there are easily distinguished cases where transport is mainly in space, and time is merely a nuisance causing a delay. There are also cases where transport is mainly in time, and an accompanying movement in space is an incidental dislocation. A reference may be made here to the bypass-germ (2) of Ex. 8.2

$$\text{pack—store—unpack.} \tag{1a}$$

If this is rewritten as

$$\text{load up—store—unload,} \tag{1b}$$

then we get something quite similar to the bypass schema (7) from Ex. 9.3, on which the battery operates

$$\text{charge—store—discharge.} \tag{1c}$$

It might be said that the battery serves to transport energy in time rather than matter or energy in space; in this respect there is here a certain analogy with the stored-up potential energy of water in the upper reservoir of Figure 1 of Ex. 9.3. Both processes operate on a bypass STS^{-1} with a passive middle part T, a feature that they share with the large class called temporary holding devices, introduced in the fifth preliminary example used in Chapter 1 and treated in more detail in Ex. 8.4.

If the battery is considered to have two states only (full and empty) then it becomes related to such standard two-state devices as on-off switches and relays and magnetic memory cores, which provide fast-memory hardware in automatic computers. The word "memory" may be noted here; we shall return to it soon. Concerning switches and relays, we might say that they work on the bypass: S (switch on)—T (keep contact)—S^{-1} (switch off). In a crude but forcible way they display the multilevel aspect of bypasses that has been already noted in many places. Here the very name "two-state device" comes from the existence of two levels or states: "on" and "off."

Since there are only two states the transformation, S (e.g., shift or switch) takes us from one state to the other, and if we repeat it again we return to the initial state. Or, S^2 is the identity transformation I: $S^2 = I$, it follows that S is its own inverse: $S^{-1} = S$. Therefore, a two-state device provides a concrete illustration of the law of excluded middle (which was briefly considered in Chapter 2). This becomes particularly clear if the physical transformation S (i.e., turning the switch or relay over) is associated with the logical operation of denial (i.e., passing over to the opposite truth-state).

Something similar happens with the magnetic core units, but here rather more will be said. It was noticed with the previous examples, and will be noticed with the examples to come, that bypass enters into transportation, whether across space or in time or in some abstract sense; we harp on this point quite intentionally. That which is transported may be matter or energy or structure; in the present instance we are concerned with information transport. Loosely speaking, information may be transported in space from point to point; the process is then called communication. Or, information may be transported in time from an instant to a later instant and the process is then called memory. Quite possibly the neutral (one is almost inclined to say "sterile") word "storage" is preferable to "memory," a term loaded with various associations. It is noteworthy that information transport, too, like matter or energy or structure transport, often depends on some form of bypass.

We shall consider the information transport in magnetic core memories, after some remarks on the early history of computing and of textile machinery. This last subject is needed here since automatic weaving is the ancestral discipline of automatic computing.

The concern here is with the idea of automatic calculating, due to Charles Babbage (1792–1871), as contrasted with the earlier notion of calculating on a machine by a human operator. Roughly put, the first stands to the second in the same relation as the automatic loom to the handloom: The former contains in one form or another the pattern of control and so it operates without human intervention; the latter is much more inert and is used by the human operator who weaves her, or his, own pattern on it. In November 1822 Babbage made an entry in an unpublished diary [25] to the effect that "... it would be extremely convenient if an (steam) engine could be contrived to execute calculations for us...." Another version, going back to 1812 or 1813, is given in a passage from Babbage's autobiography [10, p. 42] reporting a conversation with a friend: "'Well, Babbage, what are you dreaming about?' to which I replied, 'I am thinking that all these Tables (pointing to the logarithms) might be calculated by machinery.'"

The words "steam" and "machinery" deserve close attention since they provide clues to the origin of the "computer revolution," especially since they may help us to understand what delayed that revolution by approximately one century. It is well known that Babbage attempted to build for the British Admiralty a calculating machine, called by him the Difference Engine. Its purpose was to do both calculating and printing of nautical tables, so as to avoid mistakes by humans in computing and in copying and, of course, so as to speed up the whole process. Though parts of the Difference Engine were produced, the whole of it was not completed. One reason for it was that during the work Babbage has enlarged his ambition to designing and producing a complete general-purpose stored-program automatic computer, christened by him the Analytical Engine. This was to work completely automatically, on purely mechanical principles, and it turned out to be a complete failure.

Why? There are several reasons; one of them may already have been the one suggested earlier: Babbage's continual enlargement of the scope of his undertakings. So, for instance, there is a record in Reference 10 of Difference Engine 1 and Difference Engine 2. Other reasons may have been personal: He was an inventive genius of the highest order, but not a great organizer. Thus, he managed beautifully abstract ideas and concrete gears, but not collaborators, employees, or people generally. A recent biography of him [104] is fittingly titled *Irascible Genius*. Further, and probably more important, there were technical reasons. In particular, the peak of technological achievement and complexity in his day was mechanical; electrical and electromagnetic devices lay still in the future. Unfortunately, things mechanical were ill-adapted to the precision and speed required by Babbage. Using modern computing terminology, we might say that he had brilliant ideas and software but wrong hardware.

The word "automatic" comes from the Greek adjective $\alpha\upsilon\tau o\mu\alpha\tau o\varsigma$ meaning "self-acting." A computation, or for that matter any mechanical process, is called automatic if the pattern of control, that is, of correct sequencing of unit steps, is taken care of by the mechanism itself. One way to achieve it is to store in some form the pattern of control in the machine at the beginning of its operation. Now, the earliest field to undergo an extensive automation was that of textile machinery; a detailed history is given in Reference 12. Thus, as early as 1589, an English curate William Lee invented an automatic device to knit stockings. An important landmark comes in 1725 when a Frenchman, Basile Bonchon, invented the application of perforated paper for working the drawloom. This shows very clearly just how the pattern of control is stored in the automatic loom: in the form of perforations on a paper tape or paper cards. We recognize here easily the ubiquitous mark of Cain: the perforated paper card used by our computers. Bonchon's loom was modified by Falcon in 1728. Important further improvements were introduced in 1745 by Jacques Vaucanson (1709–1782), who was also one of the greatest makers of music-playing, dancing, and other performing automata [43].

However, the greatest name in automatic weaving is that of Joseph Marie Jacquard (1755–1834) to whom the invention of an automatic weaving loom is often credited; the very interesting story of Jacquard is given in Chapter 11 of Reference 12. It appears that he is given a little too much credit; a balanced estimate is contained in the following quotation [12, p. 141]: "The merit of Jacquard is not, therefore, that of an inventor, but of an experienced workman, who, by combining together the best parts of the machines of his predecessors in the same line, succeeds for the first time in obtaining an arrangement sufficiently practical to be generally employed." The first public exhibit of a complete Jacquard loom occurred in Paris in 1801.

The detailed construction and functioning of an automatic Jacquard loom are rather complicated but the principle is simply described. The art of weaving consists in taking a more or less closely spaced set of parallel-running longitudinal threads, collectively called the warp, and interlacing them with a back-and-forth running transverse thread called the weft thread. The control of the loom action consists in raising or not raising the warp threads so that the weft thread will cross any particular warp thread below or above it. Even when all the threads are the same color, one gets different sorts of weaves depending on the type of crossover. When threads of various colors are used, correct control of thread crossing allows the weaver to produce various patterns. Already with two colors, black and white, skillful control of crossover produces various shades of gray; for examples of almost photographic fidelity of such designs see the woven portraits of Jacquard and others in References 73 (p. 269) and 81. The former reference, written in 1910, contains an interesting though nontechnical comparison of the amount of information on a written page and on a piece of woven cloth.

In a hand loom the crossover is executed manually by the weaver. In a Jacquard loom it is automatically achieved as follows. A paper strip is inserted into the loom, bearing longitudinal lines corresponding to the warp threads, and transverse lines corresponding to the path of the weft. At each crossing point there is either a perforation or there is none. In case there is one, a small spindle moves through it, raising the corresponding warp thread so that the weft thread comes under. If there is no hole, then the warp thread is not raised and so the weft thread comes over. Special attention is invited to the purely binary, under-over, choice, which is the elementary control act, and to its correspondence with the on-off action of switches and relays, which was mentioned earlier in this example.

Babbage himself acknowledges his intellectual debt to Jacquard very clearly; we quote from Chapter 8 of Reference 10:

> To describe the successive improvements of the Analytical Engine would require many volumes. I only propose here to indicate a few of its more important functions, and to give to those whose minds are duly prepared for it some information.... To those who are acquainted with the principle of the Jacquard loom, and who are also familiar with analytical formulae, a general idea of the means by which the Engine executes its operations may be obtained without much difficulty.... It is known as a fact that the Jacquard loom is capable of weaving any design....
>
> The analogy of the Analytical Engine with this well-known process (i.e., Jacquard's) is nearly perfect.

We observe further that the page headings of this place in the book are "Jacquard's Loom" and "Weaving Formulae"; also, one of the two parts of the Analytical Engine was called "mill," a name with obvious associations to textile machinery. Later [10, p. 306] Babbage describes his stay at Lyons, the native city of Jacquard: "On my road to Turin I had passed a few days at Lyons, in order to examine the mill manufacture. I was specially anxious to see the loom in which that admirable specimen of fine art, the portrait of Jacquard, was woven. I passed many hours in watching its progress."

Let us return to Babbage's words "steam" and "machinery," quoted a few pages back. They point out a conceptual confusion between that which did not matter much, moving of cogs and gears and, generally, transport of motion, and that which emphatically did matter, transport and storage of information. The punched paper tapes or cards of the Jacquard loom or of Babbage's Analytical Engine constitute a sort of memory. However, punching of holes is slow and, above all, it is irreversible—we cannot unpunch a hole. On the other hand, the analogous process in magnetic core units is both fast and reversible (as well as "miniaturizable"). To sum up our sidetrackings into computing and weaving, we may say that their automatic control concerns storing, keeping, and using many times single binary-choice items of information, each of which represents either the crossover of two threads or one binary digit 0 or 1. So now we ask again: Why did Babbage fail? And we answer: One important reason was his poor implementing of the bypass (1b)

load up—store—unload

concerning the information transport of one bit of information. And why did later designers of computers succeed? One important reason was that they implemented the preceding *conceptual* bypass well, for instance, by another *physical* bypass

magnetize—store—demagnetize

which describes the operation principle of a magnetic core unit. *We see here bypass in a certain essential role that might be called wide-sense modeling.*

Since a single binary-choice item is concerned we speak here of storing one bit of information. In theory, many two-state devices could be used for this purpose, though it is clear that a switch would work badly and a relay rather better but not well either. This is because these devices are not built according to the credo: A perfect machine (unlike Babbage's Engines) has no moving parts. In this connection an incidental story concerning switches may be briefly told. Because an extensive switching apparatus forms the heart (brain?) of the telephone system, there was assembled in the late 1940's at the Bell Telephone Laboratories the famous research team of John Bardeen, Walter Brattain, and William Shockley; one of its aims was to produce a switch without moving parts. This was in fact achieved at the Bell Telephone Laboratories as the ESS (electronic switching system) but only many years later, in connection with a practical mobile telephone. What the above team did achieve was the invention of the transistor (and the Nobel Prize in physics). Thus the discovery of solid-state elements, which are superior amplifiers and rectifiers, arose from the search for a nonmechanical switch.

To return to our one bit of information, what is wanted now is an electric, electromagnetic, or solid-state device with two states, which we may call 0 and

1, or a blank and a sign. The ability to write, read, and erase fast the sole letter of this binary alphabet gives us, if we have sufficiently many copies of our device, the power to write, read, and erase any message or memorandum. Our aim can be achieved in several different ways; the one we describe relies on the physical bypass: magnetize—store—demagnetize, based on the behavior of certain ferromagnetic substances. Here the magnetic field is given by two quantities, the field strength H and the magnetic induction B. The parameters B and H are connected by a complicated relation that shows the phenomenon of hysteresis. This name, derived from a Greek verb meaning "to lag," was introduced in 1886 by J. A. Ewing [47]. Generally, two related measurable quantities X and Y show hysteresis if Y depends on X so that a cyclic change in X through its complete period causes a cyclic change in Y through *its* complete period but Y lags behind X; the graph of Y against X shows then a hysteresis loop. Hysteresis was first observed in magnetism but it has been claimed to occur in economics (e.g., when Y is the wages and X the prices) or in biomathematics (e.g., with Y referring to predator density and X to prey density).

For B and H a typical hysteresis loop is shown in the graph of Figure 1a [66, p. 80]. Let a previously unmagnetized sample of the ferromagnetic substance be placed in a magnetic field, with H rising from O to a certain value H_M. Then B rises from O to a saturation value B_S beyond which there is (practically) no further increase. The plot of B against H then gives the arc OM called the initial, or virgin, magnetization curve. On decreasing H from H_M to O, it is found that B decreases down to a positive value B_r called the remanence and giving the point P in Figure 1a. Further decrease of H from O to $-H_c$ will make B vanish, and H_c is called the coercivity. Still further decrease of H from $-H_c$ to $-H_M$ reaches the point N of negative saturation

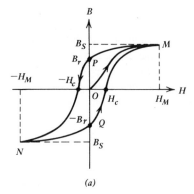

(a)

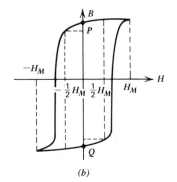

(b)

Figure 1

value $-B_S$. Now an increase of H from $-H_M$ to O takes us to the point Q with negative remanence $-B_r$. Eventually we get a closed hysteresis loop $MPNQM$, with the origin O as center of symmetry.

The shape of the hysteresis loop for different ferromagnets shows considerable variation. In particular, it is possible to produce so-called rectangular loops of the shape shown in Figure 1b. An actual memory core can now be constructed, working on the magnetize—store—demagnetize bypass. Since such cores come in large rectangular arrays, we speak of a matrix of units, and since half-pulse coincidence is used, the full name is matrix of coincident memory cores. An example of such 5×5 matrix is shown in Figure 2. There are 25 small toroidal cores, each of which receives windings from three different wires: one vertical wire $x = i$, one horizontal wire $y = j$, and a common diagonal wire R.

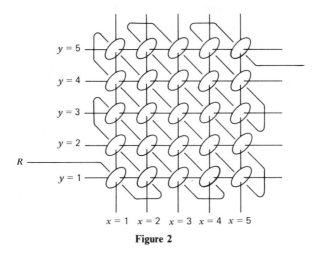

Figure 2

When no current goes through a core, its field strength H is 0, and hence its B is either $-B_r$ (called 0) or B_r (called 1). Suppose now that the core $(x = i, y = j)$ is in state 0, and we wish to write 1 on it. Let a half-pulse be sent through the wires $x = i$ and $y = j$; such a half-pulse is enough to produce the field strength $H = H_M/2$. Then in the core $(x = i, y = j)$, and only in it, the half-pulses add, and so the state 1 is produced. Suppose next that the core $(x = i, y = j)$ is to be read: We wish to know whether its state is 0 or 1. Then negative half-pulses are sent through the column $x = i$ and the row $y = j$. Again, on account of the special form of the hysteresis loop of Figure 1b, nothing happens in any core except, possibly, in the interrogated core $(x = i, y = j)$: If this one is in state 0, it stays so, if it is in state 1, it gets demagnetized to state 0. This produces a special signal in the "read" wire R

of Figure 2. It will be noticed that this readout is destructive. We do find out whether the state of $(x = i, y = j)$ is 0 or 1 by the absence or presence of signal in R; however, on our finding out, the state 1 changes to state 0. If the bit of information (i.e., state 1) is to be preserved for future use, a remagnetization is arranged, and then we get a nondestructive read-out.

EX. 9.5. SPECIAL RELATIVITY FORMALISM;
THE LORENTZ TRANSFORMATION

The classical Galilean–Keplerian–Newtonian view is associated with an objective physical nature N and (the possibility of) its (mathematical) description D(N). In the Einsteinian relativistic approach such a description is always to be referred to an observer-and-his-frame F: the description $D_F(N)$ of N relative to F replaces the universal or absolute description D(N). The basic problem that arises at once is the relation between two descriptions $D_F(N)$ and $D_G(N)$ when there are two observers F and G. Hence there appears already a certain formal similarity to Ex. 6.1 concerning the condition for two matrices to describe the same linear transformation of a vector space V, written down relative to two bases in V. In the present example N takes over the role of V and instead of two bases in V we have certain distinguished pairs of observers, or frames, say F and G. These are inertial frames that appear unaccelerated to each other; the name "inertial" is in reference to Newton's first two laws.

According to the principle of special relativity, the physical laws of nature appear the same to F and to G: $D_F(N) = D_G(N)$; therefore, the object of this example is the same as of the above-mentioned algebraic one: to assure that the two descriptions are of the same thing. But how is this sameness to be checked? In which frame, F or G? It is obvious that the two frames must be related or articulated, for comparison purposes. To begin with, the observer F has his own description $D_F(N)$. Since G is to F a part of N, F has in particular the description $D_F(G)$; we rename it as C(F,G). Here C serves as a convenient mnemonic for communication, comparability, connection, or correspondence. Of course, the reverse is also true: F is just as accessible to observation by G as G to F, and so there is the reciprocal description C(G,F) of F in the frame of G.

Now the observer F has two descriptions of N: $D_F(N)$ is the direct one, which is his own, and there is another indirect one mediated through G. To put it crudely, F commmunicates with G, then G gets his own description $D_G(N)$, and this is communicated back to F. Thus the indirect description arises from the bypass

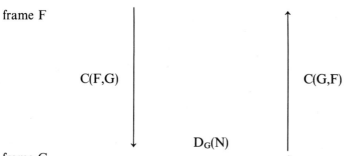

and therefore it may be schematized as

$$C(F,G)D_G(N)C(G,F). \tag{1}$$

Complete symmetry and equivalence of the two frames require

$$C(F,G)C(G,F) = I, \tag{2}$$

where I plays the role of the identity, so that in some suitable sense we should have

$$C(G,F) = C^{-1}(F,G). \tag{3}$$

Hence (1) acquires the form

$$C(F,G)D_G(N)C^{-1}(F,G),$$

and so the principle of special relativity is schematically expressible in the bypass form

$$D_F(N) = C(F,G)D_G(N)C^{-1}(F,G), \tag{4}$$

which expresses the identity of the direct and the indirect descriptions of N that F has.

The preceding schematizations leading up to (4) may be explanatory but they are loose—no meaning has yet been attached to our symbols. A justification will be given now by using suitable numbers, vectors, and transformations to describe F,G and their relation as well as $D_F(N),D_G(N)$ and their relation. As a result, (4) will lead to the Lorentz transformation.

We start with the frames F and G. For convenience, only one spatial dimension will be used: Geometrically, the frame of F is its x-axis, the frame of G is its x_1-axis, and the two axes coincide at all times. The origin O_1, of G recedes with constant velocity u from the origin O of F. Mathematically, the existence of frames F and G means a coordinatization of events in the one-dimensional space and in time; thus there exist the world vectors

$$X = \begin{bmatrix} x \\ t \end{bmatrix}, \qquad X_1 = \begin{bmatrix} x_1 \\ t_1 \end{bmatrix},$$

giving the space and time coordinates in F and in G, respectively. Physically, the existence of the frames is associated with the measurement of space and time intervals in those frames. To express the description $C(F,G)$ of G in F, the following is assumed: The length and time measurements in the two frames are connected by the transformation of world vectors

$$X_1 = LX. \tag{5}$$

Here $L = L(u)$ is a 2×2 matrix whose elements may depend on u but not on the coordinates x, x_1, t, t_1. The assumption (5) means mathematically that the transformation from X to X_1 is linear. This linearity is justified physically in the standard way on the grounds of uniformity and homogeneity of space and time.

Synchronizing the two frames in the classical Galilean manner by taking O to coincide with O_1 when $t = t_1 = 0$ leads to the Galilean matrix

$$L = \begin{bmatrix} 1 & -u \\ 0 & 1 \end{bmatrix}; \tag{6}$$

in component terms this is

$$x_1 = x - ut, \qquad t_1 = t.$$

This expresses well the kinematics of slow phenomena, which is the case of $u \ll c$ where c is the speed of light. However, the preceding description fails to correspond to experimental results for arbitrary velocities that may be sufficiently large relative to c. It is still possible to relate the frames as above, by placing O and O_1 together when $t = t_1 = 0$; this leads to the form

$$L = \begin{bmatrix} a & -au \\ b & e \end{bmatrix}, \tag{7}$$

where a, b, e are some functions of u that should reduce for u small to $1, 0, 1$ in the first approximation.

Next comes the matter of explaining the (mathematical) description $D_F(N)$ relative to a frame F with a world vector X. Let us limit ourselves to kinematics. Then, presumably, $D_F(N)$ concerns two things: extracting out of X numbers that measure quantities of physical significance to motion, and expressing laws that relate such quantities. It is well known that the simplest way of getting a number out of a vector Y is to take a suitable matrix A and to produce the quadratic form

$$Y^T A Y, \tag{8}$$

where Y^T is the transpose of Y. For purposes of illustrating some physically significant quantities of the type (8), the world vector X will be taken in the full four-dimensional form

$$X = \begin{bmatrix} x \\ y \\ z \\ t \end{bmatrix}$$

referring to the usual three spatial coordinates x, y, z and the time coordinate t, rather than in the simplified form with only one spatial coordinate x. The transpose of the column vector X is then the row vector

$$X^T = (x \ y \ z \ t).$$

Let X and X_0 be two world vectors, both of them in the frame F

$$X = \begin{bmatrix} x \\ y \\ z \\ t \end{bmatrix}, \quad X_0 = \begin{bmatrix} x_0 \\ y_0 \\ z_0 \\ t_0 \end{bmatrix}.$$

Let A, B, C, be the three matrices

$$A = \begin{bmatrix} 1 & 0 & 0 & 0 \\ 0 & 1 & 0 & 0 \\ 0 & 0 & 1 & 0 \\ 0 & 0 & 0 & 0 \end{bmatrix}, \quad B = \begin{bmatrix} 0 & 0 & 0 & 0 \\ 0 & 0 & 0 & 0 \\ 0 & 0 & 0 & 0 \\ 0 & 0 & 0 & 1 \end{bmatrix}, \quad C = \begin{bmatrix} 1 & 0 & 0 & 0 \\ 0 & 1 & 0 & 0 \\ 0 & 0 & 1 & 0 \\ 0 & 0 & 0 & -c^2 \end{bmatrix},$$

where c is the speed of light. Then the number

$$(X - X_0)^T A (X - X_0)$$

is the square of length

$$(x - x_0)^2 + (y - y_0)^2 + (z - z_0)^2.$$

the number

$$(X - X_0)^T B (X - X_0)$$

is the square $(t - t_0)^2$ of the time interval, and the number

$$(X - X_0)^T C (X - X_0)$$

is

$$(x - x_0)^2 + (y - y_0)^2 + (z - z_0)^2 - c^2(t - t_0)^2,$$

which describes the propagation of a suitable light wave. These examples go some way toward justifying the contention that $D_F(N)$ concerns forming and handling quantities of the type $Y^T A Y$, where A is a matrix and Y is a world vector or a vector simply related to it.

We return now to our two-dimensional space–time continuum, with the frames F, G and the world vectors

$$X = \begin{bmatrix} x \\ t \end{bmatrix}, \qquad X_1 = \begin{bmatrix} x_1 \\ t_1 \end{bmatrix}.$$

It is necessary to express mathematically a basic principle of special relativity:

the propagation of light appears the same in the frames F and G.

In particular, let the light wave be emitted from the common origin O, O_1 at the time $t = t_1 = 0$. Let the previously used 4×4 matrix C, which describes the propagation of light, be cut down to the 2×2 matrix

$$M = \begin{bmatrix} 1 & 0 \\ 0 & -c^2 \end{bmatrix}.$$

Then the preceding basic principle of special relativity is expressed as

$$X^T MX = X_1^T MX_1 \tag{9}$$

or in extended, though less convenient form, as

$$x^2 - c^2 t^2 = x_1^2 - c^2 t_1^2.$$

It is observed that (9) further supports our contention that $D_F(N)$ and $D_G(N)$ concern quantities of the type (8). When (5) is used and it is recalled that

$$(LX)^T = X^T L^T,$$

one obtains on substituting into (9)

$$X^T MX = X^T L^T MLX. \tag{10}$$

Let A and B be two matrices

$$A = \begin{bmatrix} a_1 & a_2 \\ a_3 & a_4 \end{bmatrix}, \qquad B = \begin{bmatrix} b_1 & b_2 \\ b_3 & b_4 \end{bmatrix},$$

it is simply checked that if for every vector X

$$X^T AX = X^T BX,$$

then we must have

$$a_1 = b_1, \qquad a_4 = b_4, \qquad a_2 + a_3 = b_2 + b_3.$$

Applying this to (10), with $A = M$ and $B = L^T ML$, we find that

$$a^2 - b^2 c^2 = 0, \qquad a^2 u + bec^2 = 0, \qquad a^2 u^2 - e^2 c^2 + c^2 = 0.$$

This system of three equations for the unknown elements a, b, e of L is solved, and we obtain

$$a = e = \pm (1 - u^2/c^2)^{-\frac{1}{2}}, \qquad b = \pm uc^{-2}(1 - u^2/c^2)^{-\frac{1}{2}}.$$

Since the Lorentz matrix $L(u)$ should reduce for small u to the Galilean matrix (6), the upper signs must be taken in the preceding solutions. Therefore, according to (7), the Lorentz matrix is obtained in its standard form

$$L(u) = \begin{bmatrix} \dfrac{1}{\sqrt{1 - u^2/c^2}} & -\dfrac{u}{\sqrt{1 - u^2/c^2}} \\[4mm] -\dfrac{u}{c^2\sqrt{1 - u^2/c^2}} & \dfrac{1}{\sqrt{1 - u^2/c^2}} \end{bmatrix}. \tag{11}$$

The complete symmetry and equivalence of the two frames F and G, schematically expressed by (2), is now verified. Since the relative velocity of G in F is u, the relative velocity of F in G is $-u$. Thus, according to (5), we should have

$$X_1 = L(u)X, \qquad X = L(-u)X_1.$$

Indeed, (11) shows that this is the case:

$$L(u)L(-u) = L(-u)L(u) = I,$$

so that $L(u)$ and $L(-u)$ are inverses.

We conclude this example with a purely speculative remark which is tempting to make but may have little or no foundation. It starts with the sort of question that might be raised by a naive newcomer to the field (or, perhaps, by an absolute master of it?): Why are there such things as c and h in physics? That is, why is there an upper bound on physical speeds and a lower bound on the limits of physical distinguishability? To ask the question in a different way: What physical consequences would have resulted in a universe with $c = \infty$ or with $h = 0$? One might guess that such consequences would be quite drastic. We recall now, by way of an analogy, the self-referential nature of the Exs. 7.1 and 7.2. In particular, we have claimed that in both cases self-reference came from a bypass the middle or the outer part of which concerned an infinity. The latter was a mathematical infinity; if c were infinite or h were zero this might have led to a physical infinity, and hence to the possibility of some drastic physical self-reference.

EX. 9.6. BYPASS AND THE QUANTUM-MECHANICAL FORMALISM

A connection between these is obvious: already the basic bypass relation $W = STS^{-1}$ happens to relate the two principal quantum-mechanical pictures, that of Heisenberg and that of Schrödinger. Let us introduce time t as parameter and suppose that the middle part T is time-invariant. With a change of scale, we get

$$W(bt) = S(t)TS^{-1}(t), \qquad b \text{ constant,} \tag{1}$$

let us differentiate this with respect to t. It is necessary to recall that the simple formula

$$(1/y)' = -y'/y^2$$

for differentiating inverses becomes in the noncommutative domain

$$(S^{-1})' = -S^{-1}S'S^{-1}.$$

Now differentiating (1) leads to

$$bW' = S'TS^{-1} - STS^{-1}S'S^{-1}$$

which may be written as

$$bW' = S'S^{-1}STS^{-1} - STS^{-1}S'S^{-1}. \tag{2}$$

Introduce the auxiliary operator $H = S'S^{-1}$, then (2) becomes

$$W' = \frac{1}{b}(HW - WH).$$

With $b = -ih$ and with H interpreted as the Hamiltonian of the system, this is the Heisenberg quantum-mechanical equation of motion [59, p. 241].

Chapter Ten

Telecommunications

Since communication may be rendered as joining, the word "telecommunication" is freely translated as joining-at-a-distance. Before the rise of science and technology such "action at a distance" was often regarded as supernatural or magic. This shows itself even in some etymologies: Telecommunication is associated with messages and messengers, in both ancient Hebrew and ancient Greek the word for "angel" means literally "messenger." Even later, during the transition from supernatural to scientific beliefs in the sixteenth and seventeenth centuries, we find such mixed natural-supernatural oddities as the statement of G. della Porta (1535–1615) in [112, Proem to Book 7]: "And to a friend that is at a far distance from us, and fast shut up in prison, we may relate our minds, which I doubt not may be done by two Mariners' compasses, having the alphabet writ about them."

This is closely echoed in the anonymously published pamphlet of an English bishop, Francis Goodwin, quoted in 1641 by John Wilkins in his own *Mercury; or the Secret and Swift Messenger* [141]: "That which first occasioned this discourse, was the reading of a little pamphlet stiled Nuntius Inanimatus, commonly ascribed to a late reverend bishop; wherein he affirms, that there are certain ways to discourse with a friend, though he were in a close(d) dungeon, in a besieged city, or a hundred miles off."

Further on, Wilkins states on his own account the following:

And it is storied of many others, that whilst they have resided in remote countries, they have known the death of their friends, even in the very hour of their departure; either by bleeding, or by dreams, or some such way of intimation. Which, though it commonly be attributed to the operation of sympathy; yet it is more probably to be ascribed unto the spirit or genius. There being a more especial acquaintance and commerce between the tutelary angels of particular friends, they are sometimes by them informed (though at great distances) of such remarkable accidents as befall one another.

But this way there is little hopes to advantage our inquiry, because it is not so easy to employ a good angel, nor safe dealing with a bad one.

We note that all the preceding schemes are pure bypasses. At the ordinary human level the communication cannot go through; therefore, it is done by a round trip bypass through some other level, for instance, the angelic level with Wilkins:

Implicit in this is the assumption that communication between angels is not subject to the ordinary laws of space, time, and attenuation.

Though Wilkins wrote on many genuinely scientific and even technological topics, and was one of the principal founders of the Royal Society of London, and with due allowance for della Porta's magnets, we are inclined to dismiss the above communication schemes as pure superstition.

In connection with the last word, there is a statement made somewhere in the correspondence of Schiller and Goethe (the exact reference could not be found but the quotation is too good to be passed up for such a trifling reason): "Superstition depends on an obscure feeling of the world's hugeness." In particular, it is felt that the world's hugeness may have room for levels other than the ordinary ones accessible to our senses. Those other levels may be either some imagined occult ones, or they may be real though hidden from our direct inspection, such as the levels of electric, electromagnetic, or atomic events. And whenever other levels exist, or are thought to exist, men, the only species that works extensively by indirection, will want to use them for bypass round trips, and especially to get closer to each other (i.e., for communication, whether by magic or by science).

Among the old methods of long-distance communication of the strictly mundane variety there is the direct transmission of written or spoken message by relays. The crudest way is to shout it from hilltop to hilltop, as was the fashion among the ancient Gauls according to Julius Caesar. The medieval Mongols organized efficient horseman relays [129], which were necessary to govern their vast empire; the same method was developed on a commercial basis in the nineteenth century United States as the so-called pony express. The Incas, who also ruled an extensive territory, employed relays of road runners who transmitted the oral message together with the accompanying quipu cord of knotted colored threads that served as mnemonic aid [94]. An occasionally used old English system was to write the message on a piece of paper or thin parchment, secure it around an arrow, and transmit this by relays of bowmen.

We note that every such method of transmission by stages comes under the schema (C_4) from Chapter 1 for bypass combination in length: The message is taken up, sent through one stage, then surrendered at the next stage, where the process will be repeated. Also, many communication nets such as those mentioned earlier are really transport nets and serve to carry objects as well as information.

The other old type of telecommunication was visual or optical; it is the "pillar of smoke by day, pillar of fire by night." Examples from the Bible and *The Iliad* are well known, to say nothing of the American Indian smoke signals and the "one if by land, two if by sea" as well as the naval system of signaling by flags and the various forms of heliography. Of course, it is clear that any optical communication method is indirect: It relies on the fundamental bypass of encode—transmit—decode. Also, it is not a goods-transport method. It might be said that two different types of nets are used for communication: transport nets and signal nets. The former transmit objects, and these objects may be made to carry information; the latter transmit information alone. However, reference is made here to Exs. 8.1. and 8.2. In Ex. 8.2 a completely imaginary "transmat" was suggested, a science-fiction contraption to transport material objects over interplanetary or inter-stellar distances through some hypothetical bypass perhaps involving laser beams and holography. If such a device could be produced then, in our present terms, a signal net would be changed into a transport net. As in Ex. 8.1, we might then run into some philosophical problems concerning the identity of that which is transmitted.

To return to optical communication, the highest development here was the system of the Frenchman, Claude Chappe (1763–1805), who appears to have coined for it the word "telegraph." Chappe's system used suitably placed towers with movable arms mounted on top; the operators had telescopes to pick up the signals from the neighboring towers. Thus the word "telegraph" is a complete misnomer; the proper name is (or should have been) "semaphore." For the history of this, and for much other early information, reference may be made to A. Still's slightly idiosyncratic book [132]. It is interesting to note how well the optical net in France functioned (aside from *The Count of Monte Cristo* type incidents): In Reference 45, a claim is made that under favorable conditions a message could be relayed across France in less than an hour. As a matter of fact, the French semaphore net may have worked too well: In the development of the telephone and telegraph France lagged behind Germany, England, and America. Incidentally, the word *telephone* or rather *Telefon*, was made up in 1860–1861 by Johann Philipp Reis (1834–1874), a German who invented a primitive telephone preceding that of Alexander Graham Bell [48, p. 65; 132]. By a process of linguistic chauvinism *Telefon* in German was naturalized to *Fernsprecher*, whereas the original form was kept in other languages.

Before taking up modern telecommunication and its relation with bypass we shall consider some aspects of communication in general. This chapter starts with the approximate equations: communication = joining, tele-communication = joining-at-a-distance; the types of communication are described and classified by specifying a number of features of such joining. For convenience, these features will be called parameters. These parameters are: distance, directionality, addressing, accessibility, equipment, channel, fidelity, capacity, and privacy, to mention some of the technically more important ones; we shall describe them shortly.

The very close connection of telecommunication and bypass is provisionally expressed by the following blanket statement. Progress in the field, whether by steady growth or by a leap, consists in improving some of our parameters, either slowly or suddenly, or in trading off the less important ones against others that matter more. Such improvement or trade-off has very often been achieved by one or more bypasses. To mention just one very general example, to which we shall return at some length, there is the important verb "to modulate." Its technical meanining is, briefly, to change a carrier signal by means of a generated input signal [34, p. 66]; however, it is a convenient catchall term referring to many types of communications signal processing before the sending. Its companion reciprocal verb "to demodulate" applies to the reverse activity at the receiving end, after the arrival of the signal. Even at this early stage, it might be guessed (correctly) that our pair of verbs refers to the outer part of a (if, indeed, not *the*) communication bypass: modulate—transmit—demodulate.

We come now to the communication parameters. The first one is distance which may be spatial or temporal, even though in the latter case a term such as recording, storing, memorializing, or even remembering may be preferred to communicating. With spatial distances a distinction of range is made, for instance, within personal presence, within earshot, within the limits of unaided or assisted visibility, beyond the horizon, arbitrary. With temporal distances there are also the grades of range between a temporary record and permanency. Of course, strictly speaking, both sorts of distance almost always occur together; at the extreme limit of defining simultaneity (or lack of it), we run here into special relativity and quantum measurement effects. An undesired temporal lag or delay occurs even in spatial cases, such as communication by modern mail. "Range" could be replaced by the more precise term "attenuation range"—the maximum distance within which communication is possible without undue decay. Some forms of communication, what we have called transport type rather than signal type, such as mail, are exempt from attenuation. On the other hand, every form of communication by signals is subject to attenuation of some form, and this leads directly to the idea of using stages with boosters (or repeaters, or relays, or regenerators). This is closely related to the previous primitive relays of horsemen, runners, or bowmen. Finally, we observe that the fundamental distinction of spatial/temporal occurs in the study of the most important method of human communication, language. The linguists divide their subject into the synchronic (or spatial) part, which views language statically or structurally, and the diachronic (or temporal) part, which is concerned with its dynamic, historical, or evolutionary development [108].

The next parameter, directionality, refers to a simple and essential division of communication: Its joining may be unidirectional (as in mail, radio, newspapers, or television) or reciprocal (as in direct speech or by telephone). Of course, a unidirectional system from A to B is in principle turned into a reciprocal one by taking two copies, one from A to B and one from B to A. However, this may introduce impracticable delays and other complications.

Since communication is a joining, this may occur between one or a few or many at one end and one or a few or many at the other end. The specification of which with which, or of how many with how many, is what we call addressing. This parameter complicates somewhat the previous one, directionality, as well as the last one, privacy. It might be claimed that much of the essence of mass communications media such as radio, newspapers, and television derives from the fact that they are unidirectional and serve for the few to address the many. For an opposite example in which the many address the few (or even one), there is that form of peaceful political protest known as the petition.

The parameter of accessibility refers to the following distinction. A particular communication system may be technically, or socially, or developmentally complex but once it is there, its use may be simple and untechnical. For example, no special skill is needed to telephone, while special technical knowledge is required to operate the telegraph. This simple binary distinction is complicated somewhat; for instance, written communication requires literacy, and even telephoning requires the recognition of numerals for dialing.

Accessibility may appear to be related to the next parameter, that of necessary equipment. This may be completely natural or highly technical, with many gradations in between. Thus a direct person-to-person communication by speech or by signs is completely natural while similar communication by telephone requires a highly technical system. It may be observed that a natural communication system is sometimes called a language. For instance, we speak of the deaf-mute language, body language, or the language of bees or other animals.

The joining that interests us as communication is chiefly the joining of humans; one may specify therefore which human effector organs are principally concerned in "sending" and which sensory ones in "reception." This is what we call channel; a more precise description of this parameter would add to "channel" the qualification "motor-sensory" or "effector-receptor." Thus the mouth, together with the rest of the speech apparatus, is the sender in direct speech or in telephone conversation, and the ear is the receiver. The hand is used for sending in telegraphy or in the deaf-mute sign language, and the eye receives. Also, the hand is used both for sending and receiving in certain touch-languages used between members of some secret groups or even between the merchants and buyers of certain objects such as precious stones. Here the last parameter of privacy is heavily concerned and the addressing is one-to-one, unlike in sign languages. The problems of sensory prosthesis, which arise with the blind or deaf people, are simply expressed in terms of this parameter: how to enable them to communicate by skillful substitution of one set of channels for another. Of course, one communicates with a blind person perfectly well by ordinary speech, but for unidirectional communication involving temporal distance, for example, books or newspapers, another channel must be substituted for the visual one. It is either the Braille print to the touch, or sound recording to the ear.

Fidelity means freedom from distortion and error, and it refers to the correctness of the communication process: That which is received should be an exact copy of that which was sent. Under favorable circumstances, certain systems of the transport type, such as mail, have an absolute fidelity, but in every communication system of the signal type there are many sources of distortion and error. There are the distortions due to the fact that the signal may have been regenerated by a number of booster stages. There are various residual impedances of the electrical circuitry elements. There is cross talk, in which signals in different but close-running leads influence each other's message (a physiological example of this arises in seasickness because certain visceral nerves run in close proximity to the vestibular ones concerned with balance regulation); in the extreme case of multiplexing the different signals may even run over the same conductor. A serious distortion cause is the presence of various fluctuating random phenomena at the atomic or electronic level in the equipment or in the transmission medium; collectively this is known as noise. Finally there are human errors due to chance, stupidity, or complex equipment. Luckily, the lack of fidelity in a communication process can be counteracted in certain cases and to some extent at least, by natural or technical overdesign. A principal factor here is called redundancy: The communication signal carries, or is made to carry, more information than is minimally required for the message.

Capacity refers to the rate at which a given communication channel can handle information; by handling we mean transmission without error. For discrete channels the information is measured in units called bits as the logarithm to the base 2 of the number of possible choices out of which one particular selection is made. This is a natural measure of information, as might already have been seen in Ex. 9.4. There the basic control act in weaving was the single under or overcrossing of two threads, that is, a single binary choice. Hence one bit of information is necessary to determine each thread crossing. The number of letters of the Latin alphabet together with the common punctuation marks is, let us say, 32 and $\log_2 32 = 5$. This means that in telegraphing by a primitive method over a cable of a number of wires, each of which could be in one out of two possible states (active or carrying a signal, passive or not carrying a signal), five wires are needed to transmit a single sign at a time. On the basis of experiments in which human subjects were made to read, type, play instruments, and so on, it is estimated that the upper bound of the human channel capacity is between 40 and 50 bits per second [110]. This upper bound is attained for reading under favorable conditions; it amounts to being able to read about one line of standard text per second (taking into account the high degree of redundancy). The preceding figure of 40–50 bits per second is picturesquely compared in Reference 110 with the capacity of a telephone channel, which is approximately a thousand times more, and with that of a television channel, which is approximately a million times more (those are 1956 figures).

The weaving control clearly refers to a discrete situation and so does reading (of printed material) since we are concerned here with recognizing a finite number of distinct signs. On the other hand, the telephone and television channels are not discrete but continuous because their signal, say the instantaneous value $f(t)$ of a current, is a continuously varying function of the time t. Since a continuous function cannot be finitely encoded, it might be thought that a continuous channel can transmit an infinite amount of information in finite time. This is not so but to explain, even superficially, what lies behind the finite capacity of a continuous channel we must go into some of the physical limitations of communication process. Consider then the preceding signal $f(t)$. As is common in communications engineering, the first step is to synthesize $f(t)$ harmonically out of various frequencies by passing from the time domain t to the frequency domain s. This amounts simply to representing $f(t)$ as a Fourier transform

$$f(t) = \int_0^W F(s) \cos 2\pi st \ ds. \tag{1}$$

An essential feature in this representation is the presence of an upper cutoff frequency W: On physical grounds, the communication system cannot sustain arbitrarily high frequencies. Mathematically, this forces $f(t)$ to be a very special type of function; by the Paley–Wiener theorem [20, p. 103], $f(t)$ is an entire function of exponential type W and is square-integrable on the real axis. A necessary and sufficient condition is that $F(s)$ from (1) be itself square-integrable but all these square-integrability conditions are easily justified on physical grounds: They mean that the total power is finite. Now, it is well known that entire functions of exponential type behave in many respects like polynomials. So, it is a fundamental property of polynomials of degree n that they are completely given by their values at any $n + 1$ equispaced points; this leads to interpolation formulas basic in numerical analysis. The same is true for entire functions of exponential type W, except that here the "degree" is so to speak infinite: Such a function is completely given by its values at a whole infinite arithmetic progression of points with common difference $1/(2W)$. The formula that results from this remarkable confluence of pure mathematics, applied mathematics, and communications engineering is known as the sampling theorem [34, p. 96]:

$$f(t) = \sum_{n=-\infty}^{\infty} f\left(\frac{n}{2W}\right) \frac{\sin\left[2\pi W(t - n/2W)\right]}{\left[2\pi W(t - n/2W)\right]}. \tag{2}$$

This does a part of our job for us; it succeeds, so to speak, in "semi-discretizing" the continuous channel: $f(t)$ is known for all t if it is just known at a certain discrete sequence of values of t. But any one of such values $f(n/2W)$ is a real number and as such it represents (theoretically only, to be sure) an infinite amount of information, for instance, as the infinite sequence of its digits. How do we perform the final step from "semidiscretizing" to actual discretizing? This is settled by recalling the previous parameter, fidelity: The presence of irremovable noise precludes complete fidelity and accuracy. In particular, there is no need to know $f(n/2W)$ exactly but only within the limits of the inescapable noise-induced error. This performs the final discretization process, and the end result is given by the famous communications-engineering formula: If the continuous channel signal has average power P, bandwidth W (as earlier), and noise (of a certain type: white Gaussian) of total power N, then the channel capacity C in bits/sec is [34, p. 326]

$$C = W \log (1 + P/N).$$

The existence of two entirely different types of communication process, continuous and discrete, calls for some final remarks; these are concerned with the fundamental CDC bypass in which the round trip is from some continuous domain to a discrete domain and then back to the continuous domain. The presence of CDC is obvious in the spatial discretization with television and in the temporal discretization with film. In both cases the return trip is accomplished as a smoothing or averaging process by the inertia of the sensory mechanisms. The most important case of CDC is probably that of nervous system functioning in organisms: The events in the environment are often of the continuous variety and so are our subjective impressions of them, but the actual neural activity that mediates the former to the latter is carried on by trains of uniform discrete pulses. This last biological bypass shows a considerable formal resemblance to a certain technological bypass essential in modern communications. The reader has probably made a telephone call recently or will make one shortly; this call starts as a continuous speech pattern and ends up as a continuous acoustic pattern. Yet the call may very well have been mediated over the wires by the PCM (pulse-code modulation) as discrete pulses (we shall discuss the PCM shortly, together with other forms of modulation).

In its highest reaches, the CDC bypass might be said to occur in the artistic process. To indicate the direction of this, we conclude the present section with some entirely subjective and highly speculative words on the subject of words. The concern will be with that perennial and ultimately perhaps unresolvable question of literary criticism: What distinguishes good literature from bad? Or rather, what distinguishes superb literature (most of Stendhal and Proust, large parts of Dostoevsky, Manzoni and Tolstoy, certain novels of Jane Austen, George Eliot's *Middlemarch*) from literature that is merely very good (Balzac, Henry James, Thomas Mann) and from literature that is bad (no examples need be named?) Presumably, every author has an ideal concept of each specific novel, and we venture to add that this ideal concept, like life itself, has the features of continuity. The medium through which the author communicates, the novel itself, is discrete, it consists of separate chapters, paragraphs, sentences, words. In the reader's mind, the reverse process may occur and the synthesis may be made from the discrete medium to something approximating the writer's continuous ideal concept. The writer's artistry may be so high that the discreteness element is obliterated without our realization, and the words melt and flow into a perfect evocation which may be sudden and swerving but by virtue of its density cannot be thin and jerky. The last two features are perhaps characteristic of bad literature, and we may associate them with the lack of the continuity of smooth flow. If the writer's control is not perfect and continuity is achieved but we are aware of the process, then the result may be what we have called "merely very good" literature. We spoke here about the art of the novel only, but it may be that something similar occurs in other arts, for instance, in music.

We come finally to the last parameter, privacy; the object here is to conceal or protect communication, and this can be done in different ways. First, the existence itself of the communicated message may be hidden by such means as invisible inks, use of microprint, surrounding the message by a large mass of irrelevant text or speech, or the ancient Spartan method of σκυταλη in which a message was written lengthwise on a strip of leather rolled aslant about a specially shaped staff and read by being wrapped on an identical staff at the destination. Such methods form the subject of steganography (this derives from Greek and means literally sheathed or enclosed writing). There are privacy systems, which might still be said to belong to steganography, where the concealment is done by a special technical trick. For instance, in radio communication the signal may be first recorded and compressed and then replayed as a brief burst, or it may be scrambled before transmission. An interesting method of concealment in radio communications was used by the Americans in the war with Japan: employing 420 Navaho Indians speaking their own (and very difficult) language [78, p. 550].

Finally, there are secrecy systems based on encoding the plaintext of the message by means of a code or cipher; this is the subject of cryptography: The content of the message though not its existence is concealed. Concerning the various privacy and concealment methods in communication, we observe only that they are often based on an obvious bypass: mask—send—unmask, compress—transmit—expand, scramble—transmit—unscramble. The clearest case is that of codes and ciphers; the bypass is then of course encode—transmit—decode, and one meets here stacked bypasses resulting from successive application of several codes. Even the terminology itself is characteristic: The branch of cryptology concerned with the first transformation of the bypass is called cryptography; that concerned with the last (inverse) transformation is known as cryptanalysis [78, p. xv].

A block diagram of a complete communication system is shown in Figure 1. Using our principal telecommunication example, the telephone, we would

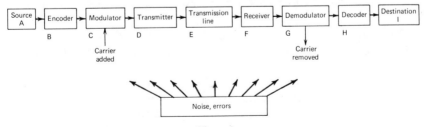

Figure 1

identify its parts as follows. The source is the person A making the call. The encoder B, sometimes also called the transducer, changes by means of a microphone device the acoustic pressure variations of A's voice into electrical current fluctuations; it encodes A's speech into electrical current. It is here in B that the essential level change of bypass, its cross-modality shift, takes place. The modulator C processes the electrical signal into a form better adapted for sending, and the processed signal then goes to the transmitter D to be sent. The transmission medium E might be a telephone cable, a waveguide, a frequency band as in wireless sending, or, as in the most modern method, a modulated laser-light beam may be sent through an optic fiber line. From here on we have the standard bypass behavior of reflection symmetry, and so the inverse operations are now executed, and in reverse order, till the call reaches its final destination, which is the person I.

A serviceable analogy using simple transport terms would be to say that the message is produced by A, the messenger appears at B, at C he is mounted on a horse or on stilts or given a motor vehicle, and at D he is sent on his way. The long stretch, the "tele" of "telecommunication," is E, the messenger's path, which also acts as center of bypass symmetry: At F the messenger is received; at G he dismounts; at H he surrenders his message to I.

In addition, Figure 1 shows noise and errors, which enter at every link and into every block though into some more heavily than into others. In similar diagrammatic representations of the communication process made elsewhere [17, p. 2; 34, p. 65; 127, p. 381], certain blocks of Figure 1 may be omitted or lumped together, depending on local need and emphasis. Also, noise and errors may be indicated as entering the transmission part only.

The diagram in Figure 1, complete of its kind though it is, has some serious faults, which arise from its horizontality: It is unduly leveling. In fact, it obliterates the levels completely. Further, it does not display in a special position the central transmission part, and it does not indicate which activities are of approximately equal status and concern the same parts of the process done directly at the sending end and in reverse at the receiving end. In brief, it does not display the bypass nature of the overall communication. Therefore, we rearrange it into the more natural form of Figure 2. This is in line with the corresponding diagrams of Chapter 8, especially those related to transport in Ex. 8.2.

Next we sketch very briefly some of the principal types and uses of modulation, starting with multiplexing. This generic term already contains the kernel of bypass in it because it refers to splicing of many signals into one at the sending end, transmitting that one, and then unsplicing the one back into the many at the receiving end. Multiplexing can be done either serially by time division, or in parallel by frequency division. To explain these two ways let us consider a very crude telephone example. Suppose that a

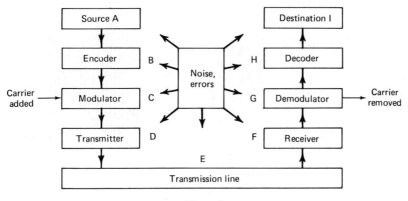

Figure 2

sufficiently faithful reproduction of voice over the telephone requires the acoustic frequency band 0–5000 Hz (hertz, or cycles per second) while the telephone cable transmits frequencies up to 15,000 Hz. It is then possible to send three separate calls together over the same cable. In the time-division method the full time is cut up into small equal intervals and each call has the whole cable during every third interval. The modulating equipment cuts up the three call signals and interleaves them into one; the demodulating equipment reverses the process.

The second method employs frequency manipulations based on the use of A.F. (audio frequency) filters, more specifically, it uses band-pass A.F. filters. Ideally, such a filter transmits freely all frequencies between certain two bounds and rejects completely all others. Now, by means of frequency shifts and band-pass filters, call 1 is assigned to the band 0–5000 Hz, call 2 to 5000–10,000 Hz, and call 3 to 10,000–15,000 Hz. The frequency-shifted signals go together over the cable, and at the receiving end they are separated and converted back into the 0–5000 Hz band. Special attention is drawn to multiplexing modulation for a laser light beam in a modern optic fiber telephone line. Here the bandwidth for a single voice channel is still about 5000 Hz, but the range of frequencies of the light beam, relative to the acoustic ones, is so enormous that a single beam can carry thousands of voice channels.

We consider next one of the oldest types of modulation: the AM (amplitude modulation); the name comes from the fact that here the amplitude of a carrier signal varies in accordance with the (instantaneous) size of the modulating signal. Only one simplified schematic illustration of this extensive type of processing will be given; it concerns the need to modulate the signal "upward" into the high-frequency region. This may be necessary, for instance, in radio transmission where the size of the antennas is approximately proportional to the wavelength and so inversely proportional to the frequency. Suppose

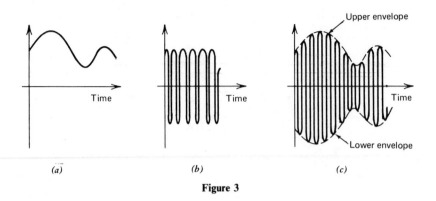

Figure 3

then that the signal produced is as shown in Figure 3a. A high-frequency sinusoidal carrier, shown in Figure 3b, is locally generated and the two are combined into the form shown in Figure 3c. This is the high-frequency, amplitude-modulated signal, which is conveniently transmitted on account of its high frequency but still carries the original message signal in the form of the envelope. At the receiving end the AM signal is fed into a suitable envelope-detector, which acts as the demodulator. This is merely a circuit with the electrical inertia large enough not to respond to the carrier's high frequency, and so the original signal is restored. Here the envelope production at the sending end, together with the envelope reduction at the receiving end, act as the outer part of a bypass; the middle part is, as usual, the signal transmission.

In the FM (frequency modulation) process the instantaneous frequency of a sinusoidal carrier is made to vary by an amount proportional to the instantaneous size of the modulating signal. The coefficient of proportionality b is called the modulation index. The mean signal power is then proportional to b^2 and the noise power is constant; thus the basic signal to noise ratio is also proportional to b^2. It is therefore possible to control it by changing b alone, without increasing signal power as would be necessary with AM modulation. Although the bandwidth of the frequencies that are required to transmit the signal also grows with b, such trade-off of improved noise performance at the cost of greater bandwidth is considered one of the principal FM advantages [34, p. 90]. Although the FM systems have more complicated circuitry than the AM ones, that important advantage was realized very early. For instance, one of the first papers on the subject, by E. H. Armstrong [5], was called "A Method of Reducing Disturbances in Radio Signaling by a System of Frequency Modulation."

A sinusoidal carrier signal is determined by its three parameters: amplitude, frequency, and phase. Varying one of them according to the instantaneous value of an input signal results in AM, FM, or ϕM (phase modulation, closely related to FM, which we have not discussed). The same procedures can be applied not only to a sinusoidal but to a pulsed carrier, of uniform rectangular pulses; we obtain then the so-called PM (pulse modulation). This has several basic advantages. In the first place, it is easy to interleave the pulse sequences resulting from PM of different signals and so obtain what has been called multiplexing by time division. In the second place, pulsed signals are better adapted to certain methods of signal generation and transmission. Finally, PM can be used like FM to obtain more favorable signal to noise ratios in exchange for more bandwidth. A description of the various PM techniques will be found in Chapter 5 of Reference 34 and in many other places.

Essentially, PM amounts to sampling the input signal, and the result is what we have called semi-discretization of the continuous input signal. The sampled values may be subjected to a quantizing process so that each sample assumes only a finite number of values. These can be numbered, for instance, as binary-scale numbers, each of which transmits easily as a sequence of 0's and 1's (i.e., on-off signals). A complete discretization has been achieved, and the resulting digital transmission scheme is known as PCM. At the receiving end the "encoding" operation of converting the sample value into 0's and 1's is inverted, and the "decoding" converts the 0's and 1's back into the sample value. Thus we have here the CDC bypass embodied in communication hardware.

It is estimated that there are some 400 million telephone sets in the world (this is the 1977 figure, from Reference 133, p. 2). The principle behind organizing telephone traffic in the large is what we have called in Ex. 8.2 the SL-bypass, and the basis on which that traffic is arranged so that a call can be made from any set to any other set is a multistage hierarchical network. This network is of the collecting-distributing type we discussed in Ex. 8.2, except that now the same units act as centers for both collecting and distributing. There may be as many as six levels or stages: individual telephones, local offices, toll offices, primary offices, sectional offices, and regional offices. An example of a call (admitted to be a rare occurrence) mediated through a full bypass stacked to the maximal depth is given in Reference 34 (p. 261).

Since the telephone traffic relies on a hierarchical-tree type of net with heavy concentration of control at the nodes, especially those of higher order, there is an obvious liability to malfunction due to failure, accident, attack, or congestion. There have been some discussions about conducting the traffic by a more homogeneous type of net with distributed rather than concentrated control [34, p. 250; 88, p. 75].

PART FOUR

SPECULATIONS AND CONCLUSIONS

Chapter Eleven

Language and Meaning

The subject of bypasses in language may be conveniently begun with some remarks on rhetoric. The relevance of bypass as a rhetorical device is forcibly shown by Tolstoy in *War and Peace* (the Maude translation, Bk. VI, Ch. VI, p. 22) in words that are, from our point of view at least, very striking; Tolstoy here describes Michael Speranski, who was for a time a favorite counsellor of the Tsar Alexander. After telling us that metaphysics was a resource the brilliant Speranski very frequently employed in argument, Tolstoy goes on to say: "He would transfer a question to metaphysical heights, pass on to definitions of space, time, and thought, and having deduced the refutation he needed, would again descend to the level of the original discussion."

This is a common rhetorical ploy of achieving one's aim by a passage to another level. In using it the arguer might even feel that it does not hurt his case at all if the bypass of the argument goes through a level that makes it rather difficult for the opponents to follow him: not impossible, nor even very hard, just rather difficult.

Concerning the beginnings of rhetoric as a subject of instruction, the title of Nietzsche's first work (*The Birth of Tragedy from the Spirit of Music*) might be parodied to "The Birth of Rhetoric from the Spirit of Litigation". According to the traditional account of Aristotle, as given in Marrou [93, pp. 84–85], the first teachers of rhetoric appeared in Sicily some time before 460 B.C. They were Corax and his pupil Tisias in Syracuse, and Empedocles in Agrigentum; the latter taught Gorgias of Leontini, who was in turn the teacher of Protagoras. Marrou writes:

Rhetoric indeed arose, not in Elis, nor even in Greece, but in Sicily. Aristotle attributed its rise to the sudden spate of proceedings for the recovery of goods that developed after the expulsion of the tyrants of the Theron dynasty at Agrigentum (471 B.C.) and those of the Hieron dynasty at Syracuse (463 B.C.), and the ensuing annulment of the confiscations which they had decreed. This helped to encourage eloquence in both politics and law, and Sicily's example is supposed to have prompted the Greeks to apply themselves with all their penetrating logic to this problem of effective speaking. Beginning with empirical facts, they gradually formulated general rules which when codified into a body of doctrine could serve as a basis for a systematic training in the art of public speaking.

Of course, rhetoric as an art to be practiced precedes considerably rhetoric as a subject codified for instruction. The ancient Greeks of the classical times realized that some Homeric heroes, such as Odysseus and Nestor, were shown as trained speakers of rhetorical skill. There is, for instance, the long and important speech of Odysseus in the second book of *Iliad*, in which he derives the prophecy of the nine years' siege of Troy from the omen of the snake snatching a sparrow with a brood of eight young birds. This speech, like some other examples of Homeric rhetoric, shows a curious bypass: Odysseus starts with the actual Achean situation in human terms and at the human level, he makes a passage to the divine level, and then, armed with suitable arguments, he returns to argue at the human level and in human terms. Such human–divine–human passage appears to be a very old rhetorical device indeed, and an extremely early example of it might be traced already in the ancient Near Eastern epic *Gilgamesh*. In the tenth tablet of the Old Babylonian version [116, p. 90], Gilgamesh is depicted grieving for his dead friend Enkidu; questing after eternal life for himself, he meets the "ale-wife" Siduri who tells him:

> Gilgamesh, whither rovest thou?
> The life thou pursuest thou shalt not find.
> When the gods created mankind,
> Death for mankind they set aside,
> Life in their own hands retaining.
> Thou, Gilgamesh, let full be thy belly,
> Make thou merry by day and night,
> Of each day make thou a feast ...

The speech of Odysseus referred to earlier, concerns prophecy and omens. This raises a general question, whether divination and prophecy may be regarded as a human–divine–human bypass. From the point of view of interest in language and meaning it might be hypothesized that interpretation of omens and oracles is an early precursor of the analysis of meaning. Fueled by the uncertainty of future, mankind's concern with omens and oracles is so very ancient that it may even reach back to the period of primitive language formation and may have influenced it. For one thing, that concern may have helped in producing the extremely important grammatical device of future tense.

It is worth recalling here that an early occurrence of the root word for our "semantics" is precisely in that context; it is a fragment of Heraclitus concerning the Delphian oracle of Apollo: "The lord whose oracle is in Delphi neither holds forth (ουτε λεγει) nor hides (ουτε κρυπτει) but intimates by signs (αλλα σημαινει)." Finally, we have that veritable catalog of the cultural and technical achievement of mankind, in the long speech of Prometheus from the drama *Prometheus Bound* of Aeschylus, one reference to which has already been made in Ex. 8.2. It may strike us as odd that such important innovations as numbers, use of metals, architecture, medicine, and domestication of animals are gone over briefly, whereas omens, divinations, and prophecy are discussed at some length. We may even speculate on what Aeschylus and other ancients would have thought about the relation of omens and oracles to language and meaning.

Another possibility for bypass to help explain rhetoric starts with the common and simple observation that certain emotional states in the speaker will modify his speech in certain definite ways. Starting with such "empirical facts," as Marrou calls them in the long passage quoted earlier, the rhetoricians employ a sort of inverse mimesis. That is, they produce suitable rhetorical conventions and, more specifically, they promote suitable rhetorical figures. These conventions and figures artificially modify the speech in the same way as the observed emotional states did naturally. The object is, of course, to help induce the desired emotional state in the audience so as to aid with persuading, swaying, or just pleasing. As Horace wrote: *"Si vis me flere, dolendum est primum ipsi tibi"*; Dryden gives this in his "Apology for Heroic Poetry and Poetic License," with the following translation: "The poet must put on the passion he endeavours to represent" (quoted by Vickers [138, p. 116]). A justification, if any were necessary, could be found in a maxim of La Rochefoucauld: *"Les passions sont les seuls orateurs qui persuadent toujours"* (passions are the sole orators that persuade always).

The inverse mimetic effect we mention here is studied by Vickers on a number of specific rhetorical figures. In Reference 138 (p. 103), he quotes Longinus, who quotes Xenophon, regarding the figure of asyndeton (meaning, literally, something untied):

In the figure of asyndeton the phrases fall unconnected as though in a torrent, almost getting ahead of the speaker himself—Xenophon writes 'With their shields striking together, they pushed, they fought, they killed, they were killed....' Since the phrases are disconnected and yet rapid, they make emphatic the excitement which both hampers the man's speech and makes it more rapid. This, the poet achieves by leaving out the connectives.

It ought to be added here that the effect may be stronger in one language than in another; for instance, in the original Greek, the asyndeton is emphasized since each one of the last four phrases of Xenophon's sentence is a single verb.

Again, concerning the figure of hyperbaton, Longinus is quoted [138, p. 104]:

This consists in a violent disruption of the natural order of words and ideas, and seems to show the most unmistakable signs of violent feeling, because those who really are angry or afraid or violently irritated or in the grip of jealousy or some other passion (for there are so many emotions that no one can hope to number them) come forward with one idea and then rush off to some other, after having thrown in something quite irrational; then they circle back to the first, and in their excitement, as though driven by an uncertain wind, shift their words and ideas now in one direction, then to the very opposite, and change their arrangement from the natural into a thousand shifting forms. The best prose writers, using hyperbaton to imitate nature, get the same effect. For art is perfect when it seems to be nature, and nature succeeds when it has art concealed within.

The phrase "then they circle back to the first" recalls the related rhetorical figures of epanalepsis (the same word or words occurring at the beginning and at the end) and antimetabole (inversion by means of the contrary); an example is the Latin proverb *"ede ut vivas non vive ut edas"* or its English translation: "Eat that you may live, do not live that you may eat." Here the bypass formula STS^{-1} can almost be used to formalize the sentence structure. Indeed, it is tempting to suggest that the rules of transformational grammar might be applied to certain selected rhetorical figures.

To sum up, the bypass under discussion starts with the observed passage from certain emotional states to certain forms of speech and ends with the return from specially arranged forms of speech to the emotional states. What comes in between is the development of some rhetorical tools and techniques. Thus the whole bypass serves as a sort of a conceptual map or, to put it simply, as a model, for certain parts of rhetoric. Some of its possible uses are obvious. A dubious though attractive application would be to attempt with it an explanation why we occasionally take an unaccountable like or dislike to a novelist, a poet, or a speaker, almost against our better judgment. The idea is simple and in spite of the appearances it is borrowed neither from

Freud nor from subliminal advertising. It may be that beside the principal overt and admitted emotions on the part of the author or speaker some other ones might be present in his production, having, as it were, seeped into it. They may produce some small but perceptible rhetorical effects to which we may be particularly sensitive on reading or hearing the text. In turn, those other emotions may be echoed in us.

We come now to a third possibility for an application of bypass to rhetoric, which will be the last one to be outlined here. It concerns a family comprehending metaphors, similies, analogies, kennings, parables, and allegories. The application will consist precisely in suggesting that it is the shared feature of bypass, which makes the items of the list a family. Our list is an extensive one; there has been much work on its members, and especially much recent work on metaphor, including several interdisciplinary symposia. The interest comes from many directions: literary, scientific, philosophical, linguistic, theological and, occasionally, even commonsensical. There is much material here so that we shall be forced to be very sketchy and highly selective in order to stay within our narrow compass.

We start with kennings, which are special metaphors from Norse and Anglo-Saxon poetry. They are extremely brief, being compounds of two or (rarely) more words, usually nouns, as in the Old English "mere-hengest" or "sea-stallion," which stands for ship. Other examples are "ring-giver" for lord, "battle-gleam" for sword, "battle-adder" for arrow, a collection of circumlocutions for sea: "whale-road," gannet-bath," "swan-path," "water-street," and "wind-domain." The entry on kennings in Reference 113 (p. 434) cautions us that they are "portmanteau devices whose suggestive associations deserve to be unpacked with care." A good example is the rather nontypical kenning "peace-weaver" for wife. Its basis is not, as might be assumed, that women weave while men fight and that weaving is an epitome of peace. This is wrong because it is indeed (assumed that it is) women who weave, but "peace-weaver" stands for "wife," not for "woman." It is simply that a marriage often ended a feud between two families or tribes. Thus our kenning turns on an analogy that assimilates the orderly weaving together of two cross-running groups of threads to the peaceful bringing together of two once-warring groups of men.

The two-word brevity of kennings may make it easy to abstract some general features of metaphors. In the first place, there are some two domains that need not appear by name; then, there are two individual terms, one associated with each domain, whose relation is the pivot of the kenning, or the metaphor in general. The domains may be land, sea, air, other subdivisions of nature, a craft, a segment of society, a type of activity, and so on. With "sea-stallion," one domain is explicitly named as sea and the other is implied to be land; the individual terms are ship, which is what the whole kenning stands for, and stallion. With "battle-adder" one domain is battle and the other one may be conjectured to be nature; the individuals are arrow and adder. The whole kenning or metaphor may be sometimes reported as an analogy and occasionally even as a proportional analogy. Thus the last two examples lead to

$$\text{ship}:\text{sea}::\text{stallion}:\text{land},$$
$$\text{arrow}:\text{battle}::\text{adder}:\text{nature}.$$

This has the mildly amusing effect of producing "dual" kennings: "land-ship" for stallion, and "nature-arrow" for adder. The first one is obviously related to such actual metaphors as "desert-ship," "prairie-schooner," and "land-yacht."

The two-domains structure that goes with metaphors might already be deduced from the etymology of the word "metaphor," which, as mentioned earlier, literally means "transfer." The etymological element continues and the two-domains aspect is emphasized when we come to allegory. For this comes from two Greek words αλλος ("other") and αγορευειν ("to speak"), giving us something like "other speech," in the sense of speaking of one thing in terms of the other. This, of course, reinforces the common observation that allegory is a sustained or continued metaphor, as well as Aristotle's observation that "metaphor is the substitution of the name of something else." In his article on allegory [113, p. 12], Northrop Frye says: "We have allegory when the events of a narrative obviously and continuously refer to another simultaneous structure of events or ideas, whether historical events, moral or philosophical ideas, or natural phenomena." And later, he explains: "Allegory is thus not the name of a form or a genre but of a structural principle in fiction." We venture to add that the essence of Frye's principle is a correspondence between the two structures; a comparison might even be made here with the correspondence between mathematical structures in Ex. 6.4. At any rate, in terms of our bypass model, it might be said that a metaphor is a single round trip ticket between two levels, whereas an allegory confers the freedom of unlimited travel between them.

No comments are needed on the subject of parables since these are special allegories, short and reasonably transparent, in which the "first structure" narrative is simple and the "other" structure is moral. Their relative abundance in the New Testament may be one of the reasons for the modern Christian theological interest in language, especially in its metaphorical uses.

The occurrence on our list of both similes and metaphors recalls the delicate questions of their relation and convertibility. Answering these questions probably would require a searching analysis of the relation between the two individuals on the two levels, and various candidates have been suggested for that relation: analogy, assimilation, collision, comparison, confrontation, contrast, fusion, identity, juxtaposition, parataxis, similarity, tension. Luckily, the proposed bypass model is neutral in that it does not commit us for or against any of those suggestions. On the whole, the modern opinion favors stricter separation between simile and metaphor; the older and classical view is for closeness. For instance, Priestley [114, p. 181] says: "A Metaphor hath already been defined to be a simile contracted to its smallest dimensions." Aristotle in his *Rhetoric* recommends that "when the metaphor seems daring, let it for greater security be converted into a simile," an opinion which condenses facetiously to: A simile is an insured metaphor. With our model there is considerably more point to describing a metaphor (or a simile) by a simile: It is like the voyage of a Columbus. That is, there is a return, and there may be profit and glory, but these are not guaranteed. The further the separation of the levels, the longer the voyage and, presumably, the greater both the profit and the glory. This paraphrases Quintilian's "The more remote the simile is from the subject to which it is applied, the greater will be the impression of novelty and the unexpected which it produces" and also (though not as well) Dr. Johnson's "A simile may be compared to lines converging at a point, and is more excellent as the lines approach from a greater distance" [both quoted in Reference 113 (p. 768)]. Finally, our simile may be used to explain how similes and metaphors get worn out: When the voyage has been replaced by an air return trip booked with a travel agent, some profit may remain but the glory is gone.

There remains the knotty subject of analogies, and a few opinions of scholars past and present will be collected first, concerning analogy and its relation to metaphor. Lloyd [90, p. 175] says:

Here I shall take analogy in its broadest sense, to refer not merely to proportional analogy (a : b :: c : d) but to any mode of reasoning in which one object
or complex of objects is likened or assimilated to another (of the two particular
instances between which a resemblance is apprehended or suggested, one is
generally unknown or incompletely known, while the other is, or is assumed
to be, better known)....

Later (p. 176), he adds: "What I term analogy is, of course, an extremely general mode of reasoning, and one common in some form to all peoples at all periods of time."

On analogy and metaphor, Palmer [106, p. 315] writes: "To achieve this, linguistic symbolism turns to its service what is perhaps the most remarkable quality of the human mind—its capacity for analogy: that is the ability to perceive similarity of quality or relationship in dissimilar objects or situations. This constitutes the mental basis of what is known as metaphor." Much briefer is the amusingly disarming statement made by Anttila [4, p. 141]: "Metaphor is an important kind of analogy or vice versa. . . ." Already Aristotle in classifying metaphors into four kinds in *Poetics* (from line 1457 on) said: "Metaphor is the substitution of the name of something else, and this may take place from genus to species, or from species to genus, or from species to species, or according to proportion." The last word reminds us that in Greek *analogia* means proportion, so that the last phrase could be translated "or by analogy." An important element of Aristotle's quotation may be rephrased in modern mathematical terms: The substitution that constitutes the metaphor concerns either class membership or analogy. Finally, we repeat the quotation from Jevons referred to in Chapter 2: "It has been said, indeed, that analogy denotes a resemblance not between things but between the relations of things."

An example supporting Jevons' claim is in the following quotation from Chapter 12 of the Dumas romance *Twenty Years Later*; it serves also as an instance on which to mark some details of the bypass passage. D'Artagnan is looking for his friend Porthos who has come into some landed property; on his way there d'Artagnan sees from a distance the towers of a "feudal" castle, and then an elegant though less impressive palace built later: "The road led in a straight line to this handsome palace which, compared with its ancestor, the Castle de la Montagne, was like a fop in Duc d'Enghien's train by the side of a barbed knight of the times of Charles VII." On account of its "like" it is formally a simile, two metaphors are buried in it (the palace is a fop; the castle is a knight), and it obviously expresses an analogy and even a proportional analogy (though of course not a numerical one) of the form

$$\text{palace} : \text{castle} :: \text{fop} : \text{knight}.$$

It is interesting to note that the domains, or levels, can be arranged in two different ways so that two essentially different bypasses arise. One is of the form STS^{-1} with the schema

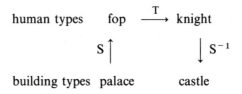

where S is the cross-type matching of human to building and S^{-1} the inverse cross-type matching, and T is the level matching of individuals of the same type. However, the context indicates that Dumas had in mind the analogy expressed by the other bypass, with the schema

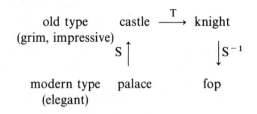

This emphasizes the distinctions between the old and the modern, so dear to the heart of Dumas who rather fancied himself a historian. Now S is the relation of like things (palace to castle, fop to knight) and T is the analogy relation, so that indeed the analogy denotes a resemblance between relations, as Jevons would have it. The bypass reads: Palace in relation to castle is as knight in inverse relation to fop. We note one peculiarity: The order of words in the quotation is ... palace ... castle ... fop ... knight; according to the arrows in the schema this is taken care of by the inverse S^{-1}, which interchanges "knight" and "fop." It will be recalled that exactly the same peculiarity of sequencing was met in Ex. 4.0, concerning numerical proportion.

So, to continue, is analogy a relation of two or of four things? According to Jevons analogy is a relation of resemblance, hence it involves two terms (or perhaps more) but each of those terms is itself a relation. Assuming the latter to be at least binary, we have four terms (at least) entering the analogy. On the other hand, Lloyd claims in analogy a resemblance between two, not four, terms. Using our model, the following reconciliation of this apparent contradiction may be offered. Consider a simple analogy with two terms in it, of the simile form: a is like b. Even here each individual term resides in some totality, or refers to some domain, say, a goes with A and b with B. Then "a is like b" is only a short two-term form of the four-term analogy that may be reported as "a is in A as b is in B" or "a is to A what b is to B"; there is another four-term form with a stronger proportionality suggestion: "a to A is as b to B." The same suppression of some of the containing or reference domains was already seen with the kennings and metaphors generally. We note that a simple two-term metaphor, such as "Achilles is a lion," is also a contraction of the full four-term form "Achilles (among Acheans) is (like) a lion (among animals)."

An analogy may also be concerned with class characterization rather than with class membership; an example is the often made analogy of pitch to color. This means of course that pitch is to sound what color is to light, and so the full four-term form here would be: Pitch is in the acoustical domain what color is in the visual one.

It might be thought that analogies based on class membership or on class characterization are something entirely different from the proportional analogies of the type a:b::c:d. This brings us to our principal problem: What is the common purpose, if any, of the two types of analogies? For that matter, what is the common purpose, if any, of analogies, similies, metaphors, kennings, allegories, parables? We shall claim that there *is* one and that it may be traced by starting precisely with the proportional analogy, where the matters are simple (perhaps even too simple). This takes us back to Thales and his work on similarity, the theory of proportionals, and the measurement of the height of an Egyptian pyramid (considered already in Ex. 4.0 at a length that may have seemed excessive).

As a starter we recall that Lloyd in the preceding quotation is anxious to have us note how one of the two terms in his analogy is unknown or incompletely known. His emphasis makes all the more curious the omission from his book of any reference to Thales' work on proportionality. This work of Thales had a simple practical purpose: the measurement of inaccessible magnitudes. Expressed in more detail, its practical aim was the measurement of magnitudes which are open to some observation but, like the height of a pyramid, are not amenable to simple and direct measurement of such primitive type as by a stretched rope. The key terms are here "simple" and "direct": Although no simple direct measurement is possible, a complex indirect one may be possible. What the bypass of proportion achieves is to overcome a spatial inaccessibility by indirection in order to measure something. We claim that, with some changes, that is also what bypass achieves in analogies, similes, metaphors, kennings, allegories, and parables. The inaccessibility is now not spatial but of a different, perhaps subtler, nature, and the aim is not to measure but to describe, explain, or enhance. The complexity and the indirection are as before, in proportionality; so is the main purpose: overcoming distance or, as usual with bypass, transport. This may perhaps suggest the appositeness of the verse of Rilke that opens this book.

The next topic we briefly take up concerns the proper parsing or segmentation of spoken or written language that must be done if the intended meaning is to be extracted. To see what is meant and what may go wrong, even with units as large as whole sentences, let us consider two examples. The first one is from an apocryphal story about a prisoner in late Tsarist Russia who had some friends at court. These tried to get him pardoned by submitting new evidence which, however, turned out to be so officially damning that the following directive was sent by telegraph to the prison governor: "Pardon impossible. To execute immediately." Through some providential mistakes, which might even have been due to an obliging fly, the message arrived as: "Pardon. Impossible to execute immediately." It seems that the prisoner lived.

The other example is apparently historical and concerns the murder of

King Edward II of England. The following unparsed Latin message was produced by an influential English bishop, Adam Orleton [119, p. 97] for Queen Isabella, Edward's wife, and sent to Edward's jailers:

Edvardum occidere nolite timere bonum est.

This obviously consists of two distinct clauses that must be obtained first, and it does not matter whether a comma, semicolon, or even a period is used to separate them. If the parsing sign occurs between *nolite* and *timere*, then the sense of the message is:

You do not want to kill Edward. It is good to be afraid.

But if the parsing sign is inserted between *timere* and *bonum*, then the sense is almost exactly opposite:

You need not fear to kill Edward. It is right.

Our examples are extreme to the point of gruesomeness so as to stress the importance of correct parsing. Of course, there are some obvious devices to help with parsing or at least with partial parsing: punctuation marks and the use of capitals in writing, longer and shorter pauses, tonal changes, and even gestures, in speech. Before connecting the subject of parsing with bypasses it will be necessary to complicate it in one realistic and important way: by bringing in discontinuous segments in addition to continuous ones. Our two examples show ambiguity arising from different parsings of the same strings of words:

$$\text{Ex. 1: } (X_1 X_2)(X_3 X_4 X_5) \quad \text{or} \quad (X_1)(X_2 X_3 X_4 X_5),$$

$$\text{Ex. 2: } (Y_1 Y_2 Y_3)(Y_4 Y_5 Y_6) \quad \text{or} \quad (Y_1 Y_2 Y_3 Y_4)(Y_5 Y_6).$$

Every segment in both examples is continuous: It consists of consecutive words. However, there occurs also another type of segment, discontinuous one: This consists of words separated by other words that do not belong to it. A simple example is provided by the sentence

John, though poor, is honest

with the obvious segmentation shown already by the commas

Here we have the core sentence $X_1 X_4 X_5$ (John is honest), which is a discontinuous segment, and a qualifying phrase $X_2 X_3$ (though poor), which is a continuous segment. An even more complicated case occurs with the brief French sentence without any internal punctuation

Il les a battus (he has beaten them)

in which the agreement of the singular and the plural forms suggests two discontinuous segments

$$\text{(} Z_1 \diagup Z_2 \diagdown Z_3 \text{)} \diagup Z_4 \text{)} \qquad . \tag{1}$$

We quote next the famous ambiguous example

Flying planes can be dangerous

(with the parenthetical observation that those who are tired of meeting this again and again may replace it by the following sentence of completely isomorphic structure: Frying friends may do wonders). The ambiguity of meaning arises from the defective English verb "can" which has the same singular and plural forms: If the potential danger is replaced by an actual one then in one meaning "can be" is replaced by "are," in the other by "is." Also, in the first case the subject is the plural noun "planes," and in the second one it is the singular gerund "flying." Probably the simplest way to distinguish the two meanings is to observe that the first one comes from the continuous segmentation

$$(X_1 X_2)(X_3 X_4 X_5),$$

and the second one from the discontinuous segmentation

$$\text{(} X_1 \diagup (X_2) \diagdown X_3 X_4 X_5 \text{)} \tag{2}$$

Now there arises the problem of producing in a simple and useful way a segmentation of any string $X_1 X_2 \ldots X_n$. If there are no discontinuous segments then the ordinary brackets do the job. But with the discontinuous segments this does not work: If (2) were bracketed as

$$(X_1 (X_2) X_3 X_4 X_5)$$

then (1) would have to appear as

$$(Z_1(Z_2Z_3)Z_4)$$

which is wrong. Numbering the corresponding left and right brackets thus

$$(Z_1(Z_2Z_3)Z_4)$$
$$1 \quad 2 \quad\quad 1 \quad 2$$

does not work either. We now propose a way of solving the segmentation problem, which is based on two bypass ideas. The first one is derived from Ex. 8.4 which dealt with various "temporary holding devices"; the first stated purpose of such devices was temporary positioning. One of the principal instances of temporary holding devices was scaffoldings, in their role as aids to the erection of structures. The other idea is supplied by the word "level," which has occurred so often in connection with bypasses. Putting together the two ideas, we shall display the elements of our string $X_1X_2 \ldots X_n$ on various levels, placing a separate segment on each level. Thus (1) becomes

$$Il \quad\quad a$$
$$les \quad battus$$

or

$$Z_1 \quad\quad Z_3 \quad .$$
$$Z_2 \quad\quad Z_4$$

This still suffers from the inconvenience of not following a lineal presentation. Therefore, we bring in the bypass terminology and rewrite the preceding schema on-line as

$$Z_1SZ_2S^{-1}Z_3SZ_4S^{-1}.$$

Here S means descend one level and the inverse, S^{-1}, means ascend one level so that the terminal S^{-1} in the string returns us to the line. It is clear that any string can be segmented in any way whatever. First we write it on as many levels as there are segments, with each level containing just one segment. Then this is rewritten in the on-line fashion by repeated use of the operators S and S^{-1}.

Some combinatorial problems turn up now; for instance, we may ask: What is the number B_n of different segmentations of a string $X_1X_2 \ldots X_n$, not counting as distinct those that differ only by a permutation of levels?

It is shown in combinatorial literature that the quantities B_n, called the Bell numbers or the set-partition numbers, are given by

$$B_n = \frac{1}{e} \sum_{k=0}^{\infty} \frac{k^n}{k!}.$$

No further reference is made to this aspect beyond remarking that some new combinatorial questions could be asked now (e.g., concerning the algorithms for detecting various types of discontinuous segments, the numbers of corresponding segmentations, etc.).

So far, the levels refer to distinctness and not to importance: The segments at the lower levels are in no sense more or less basic than the others. However, this may be changed, with a possible application to a type of preprocessing that precedes the extraction of meaning. Our sketch considered only partial segmentations rather than complete ones. A complete segmentation, down to the limit of words, can plainly show more than a partial one. In particular, it can show something of the genesis or history of a sentence. Suppose, for instance, that the sentence "flying planes can be dangerous" is assigned the following history:

> flying
> flying can be
> flying can be dangerous
> flying planes can be dangerous.   (3)

Other histories, of course, are possible; the preceding one, with the second meaning in mind, could be defended on the grounds of starting with the subject, then adding the compound verb, and then expanding by further addition of predicate, object, and so on.

How would we represent the above history as a complete segmentation? In the first place, (3) must be turned upside down. For, the hearer or the reader is presented with the complete sentence only, the finished product and not its history. If (3) is assumed to reflect, however crudely, the process of producing the sentence, then presumably the process of understanding it may well involve a reversal of the production process. In fairness, it must be admitted that this smuggles in our basic hypothesis of language as bypass; it is precisely the reversal symmetry, so characteristic of bypass in general, that suggests the reversal here. Reversing the list (3), we get

> flying planes can be dangerous
> flying   can be dangerous
> flying   can be
> flying

Only the bottom word of each vertical column is needed for complete information, and so the preceding set-up becomes

$$\text{planes}$$
$$\text{dangerous}$$
$$\text{can be}$$
$$\text{flying}$$

In our on-line notation this is

$$\text{SSS flying } S^{-1}S^{-1}S^{-1} \text{ planes SS can be } S^{-1} \text{ dangerous } S^{-1},$$

and now, not surprisingly, the subject "flying" appears at the deepest level, the compound verb at the next deepest, and so on. It is tempting to say that the level allocation reflects now something like the relative importance in the sentence of the constituent words. Finally, suppose that such a complete segmentation of a sentence leads to a history that produces either no meaning or suspect meaning. It might still be possible to fix things up by seeing what happens when the segmentation is changed.

Incorporating the problem of meaning into the general problem of language, we ask now "what is language?" and we add at once that a complete or even a substantial answer to that question would tell us what it is to be human; such an answer therefore is not to be expected. But in the previous chapters, and in the preceding parts of this one, the hypothesis was made, surreptitiously several times and openly once or twice, that language is a bypass (thoughts to words—transmission of words—words to thoughts). Thus it might appear that we do both: deny the possibility of answering the question and answer it ourselves. Therefore, it must be explained that our hypothesis "language is bypass" is not a complete or a substantial, or even a partial, answer to our question. But then, what is it, and why make it? To put it very briefly, it is a guide to a possible model that might help to rephrase and partially answer the question "what is language?" This must be followed a little further, partly for its own sake and partly to underscore the role of the bypass principle as a model for making models.

Some general comments will be made first. Let us compare the following progressively more ambitious statements concerning the occurrence of bypasses:

1. Matrix similarity is a bypass.
2. Method of integral transforms is a bypass.
3. Telephone communication is a bypass.
4. Language is a bypass.

Statement 1 is from Ex. 6.1. Statement 2 is from Ex. 5.9. Statement 3 is the second example in Chapter 1 and it also appears at length in Chapter 10. Statement 4 is our hypothesis that language, too, is an instance of bypass.

There can be no argument about (1). With the minimum of technical knowledge the equation $W = STS^{-1}$, defining the similarity of matrices W and T, is proved to be the necessary and sufficient condition for W and T to express the same linear transformation. No question arises about the sense in which S^{-1} is inverse to S, nor could we ask whether S and S^{-1} involve similar or different techniques or hardware. The bypass round trip results in exact return, everything is formal and rigorous though perhaps unenlightening (from the point of view of bypass modeling, that is).

The matters are different with (2). Here $W = STS^{-1}$ is no longer a formal equation as in (1), but a symbolic relation asserting the conjugacy of the process W of solving a harder problem to the process T of solving an easier one. S denotes taking the transform and S^{-1} taking the inverse transform in order to "return." The relation $W = STS^{-1}$ is the complexity reduction schema: Direct solution W being impossible or beyond our powers, an indirect three-part bypass method is used. We can ask about the sense in which S and S^{-1} are inverses, and the principal theorem for each kind of transform (Fourier, Laplace, Mellin, etc.), which is called the inversion theorem, answers that. However, unlike in (1), the round trip return is not exact: $S^{-1}S$ or SS^{-1} applied to a function f results in a function that may differ from f (though only inessentially—over a set of measure zero). It makes sense to ask whether S, T, or S^{-1} is the hard part (it is most often the "return" S^{-1}) and whether the techniques for taking S and S^{-1} are similar (they are not). Something important may be noted. In passing from (1) to (2), there is a loss of formality and rigor but also a certain gain: The bypass model begins to explain things with integral transforms, whereas it does not do so in the case of matrix similarity.

On coming to (3), a claim of novelty may be made. For the case 2, the interlevel round trip idea at least, though of course without bypass terminology, is nothing new. It occurs in the diagram form in classical texts such as Reference 41 (pp. 347, 397) to explain the integral transform technique. But it seems that nobody has previously considered (1), (2), and (3) too, to say nothing of further bypass examples, as manifestations of a single principle. On first hearing (3), after an introduction involving (1) and (2), people showed four different types of reaction. At one pole there was the enthusiastic acceptance with an expression of pleased insight; at the other extreme, there was the attitude of "So what?" or "So what else is new?" obviously associated with "There is nothing in it." Nothing further need be said about those two extremes.

It was the reactions of the two intermediate groups that were of interest.

On the one hand, there was a qualified though critical (and sometimes highly critical) acceptance; on the other, a qualified skepticism but with an open mind. It was in the highly profitable discussions with those two groups that various questions arose concerning the more precise interpretation of the bypass equation $W = STS^{-1}$ in communication-theoretic terms, the sense in which S and S^{-1} are inverses, the possibly different hardware and software of implementing the transformations S and S^{-1}, the possibility of recasting some of communications theory in bypass terms (e.g., for instruction purposes), and so on. Both groups were favorable to some developments of Chapter 10 treating (3) in more detail, and especially to the repeated appearance of reversal symmetry seen in bypass extension in depth, which is what we call bypass stacking. In fact, once convinced or self-convinced, even partially, the critical supporters contributed many examples and details concerning bypass in communications. On the balance, the opinion of those in communications was in favor of accepting (3), minimally as a device of methodological use, possibly as an instructional technique, and perhaps (though with considerable reservations and qualifications) as a guide to research and innovation.

Passing from (1) to (2), then to (3), and finally to (4), one meets some gradual changes as well as sudden jumps. On the gradual side the formal rigor of the bypass equation $W = STS^{-1}$ decreases steadily while the explanatory possibilities of the bypass model grow (as we fondly imagine). As to jumps, the subjects in (1) and (2) are remote and impersonal to us, unless perhaps we happen to be mathematicians engaged in, say, matrix inversion or Laplace transform. But (3) is no longer remote and impersonal since each one of us may well be involved at one end of a telephone connection. Still, the subject matter of (3) is in the obvious sense public and exterior. On the other hand, with (4) this is no longer so; the outer part $S \ldots S^{-1}$ of the language bypass is interior: The transformation S from thoughts, wishes, concepts, etc. to words occurs somewhere deep within our minds, and so does the inverse transformation S^{-1}. It is only the transmission T of words, together with the two "interfaces," vocalization and hearing, that are exterior, public, directly accessible to observation. So, it would seem that we land in some subjective morass of introspection when it comes to the outer part $S \ldots S^{-1}$ of the bypass in (4).

This puts a considerable premium on producing first a reasonably acceptable bypass of the "mental" type, concerned in part at least with observable behavior. In our first example in Chapter 1, we have considered different animal reactions to an obstacle placed on the way to food. Several questions were raised then, the last one being, "Is there any connection, formal or otherwise, between bypass and such human development theories as that of Piaget?" As is well known, Piaget advocates a theory in which infants and children develop by many stages. On the purely formal side, the progress from some of Piaget's stages to the higher ones looks as though the children were learning bypass stacking to greater depth. However, there is a more definite instance concerning infantile motor behavior; unfortunately, the author cannot recall where in the huge output of Piaget and collaborators the following observations were made. Piaget notices repeatedly that before movement has been well learned the infant will execute "an empty run" sequence: move a hand forward, as though to grasp something (and take it), then move it back. This "empty run" looks like the outer part SS^{-1} alone of a bypass, before the infant has learned it well enough to interpolate some useful middle part T in between S and S^{-1}.

Still, there seems to be no evidence that children learn the use of language in this way, that is, by practicing empty runs SS^{-1} on the language bypass STS^{-1}, before they have learned the middle part T which is the transmission of articulated sounds. The empty run SS^{-1} would here amount to something like hearing oneself think. The difficulty of explaining how language is acquired, both individually (by children) and collectively (by man as species), is a considerable part of the difficulty of the problem of language. A recent book on the history of the problem of the origin of language [131] lists many of the explanations that have been attempted and of successive objections to them. One difficulty appears to have been often raised; it seems to concern a form of infinite regress and is: in order to learn our first language we need to know a language to learn that first language with.

But what about the possibility that the first language we learn is already our second language? Is there not some inner working code of the brain itself? It may be that beside, or rather underneath, the learned structure of the natural language (English, Chinese, etc.) that we think we think in, there is another structure, which is innate and not learned, a yet more basic code. The new hypothesis concerns obscure, not to say profound, matters and the usefulness of our bypass-modeling terminology may be observed already in the ease of formulating this new hypothesis: The simple bypass STS^{-1} is only the middle part of another bypass that is stacked.

Expressed at greater length, this is as follows. The hypothesized bypass $W = STS^{-1}$ of (4) provides schematically for communication W from the private thought-level of one individual X to the private thought-level of another individual Y. But direct communication W is impossible, unless one grants something like telepathy. So, the communication occurs by the usual bypass round trip. X first verbalizes his thoughts by translating them into a pattern of articulated sound, this transformation S is followed by the transmission T through the public verbal level from X to Y, and Y then executes the inverse transformation S^{-1}. Under the new hypothesis the whole process is augmented by supposing that both individuals X and Y have something that may be called (private) basic level, in default of a better word. Now the essential communication between X and Y is by a stacked bypass $USTS^{-1}U^{-1}$ involving the private basic level, the private thought level, and, in the middle, the public verbal level.

Very likely it will be thought that the above augmented hypothesis amounts to obscuring what is already midnight darkness by putting inside a black box another blacker box. However, we shall submit now some evidence, admittedly very partial, indirect and inconclusive, toward our new hypothesis. This evidence is based on data privately collected over many years concerning the age at which children, who *will* become mathematicians or serious users of mathematics, learn to speak. Statistically, these data indicate a substantial positive correlation between future mathematical ability and relatively late age of learning to speak. The outstanding case is that of Albert Einstein, about whom it is known that he learned to speak at an age closer to three years than to two.

How does the substantial positive correlation claimed to occur between future mathematical ability and late onset of speech support our hypothesis of two interior levels? The answer is best given in terms of a fragmentary ad hoc model, and this model is most easily described by a moderate use of computer terminology.

Modern computing machines have separate hardware parts, such as the input equipment which accepts the problems, memory units which store information items, the central processor unit which executes the basic logical and arithmetical operations (the latter by reducing them to the former), and the output equipment which brings forth the answers. These parts operate in different "languages." For instance, the input equipment may accept problems stated in the so-called FORTRAN language (FORmula TRANslation). This is a reasonable facsimile of the standard mathematical language of elementary arithmetic, algebra, and analysis. Or a more modern version, the so-called PASCAL language, may be used. In particular, numerical data are presented in ordinary decimal form. Such problems are then translated by the machine itself into its own on-off binary type of language. This is akin to the Morse code, Boolean switching notation, and even the on-off neurophysiology of nerve fibers. Also, it may recall the basic operation of the Jacquard loom, described in an earlier chapter. This inner machine language is more suitable for the central processor use. The answer to the problem is obtained by the central processor, it is then retranslated (e.g., the numerical data are converted from the binary into the decimal form), and printed out by the output equipment.

There may be other intermediate levels, between the peripheral and the central, with their own specific languages. For our purposes it is enough to consider the programming language which may be called I/O-language (for input-output), and the machine language which may be called C-language (for central processor). Roughly speaking, the I/O-language is for communication to and from the machine, the C-language is the one in which the machine itself, or rather its central processor, works. The translation from the one to the other is done by means of the so-called assemblers and compilers. The reason for such complicated arrangements is the following. In terms of the intrinsic functioning of the machine itself, its own machine language, there would exist an almost impassable barrier between it and the user. That is, the amount of minute practical detail, and the attention required, put it beyond most computer users to present their problems to the machine in its own language. We may even add, though incidentally, that the complicated arrangements involving more than one "language," referred to earlier, amount to this: The communication gap between machine and user is bridged by a bypass.

It may be hypothesized now that the ordinary spoken languages that children learn correspond to the I/O-language of the machine while mathematical reasoning and problem solving may be done in the brain in terms of some C-language. It would appear that, other things being equal and good general intelligence being granted, there are two possibilities:

1. The child who learns to speak late has a better chance to develop his "central processor," due perhaps to linguistic insulation from the environment, or
2. Conversely, the child whose "central processor" activity starts earlier is less likely to begin learning the I/O-language right away, on account of more intense inner preoccupation.

A claim has been made that the "basic-level code" in humans is innate, whereas the natural languages are learned. Since the computing machines are man-made artefacts, it would make no sense to claim that their C-languages are "innate" while the I/O-languages are "learned." But it may still be granted that the C-languages are less arbitrary than the I/O-languages and show considerably less variety. In some sense, they are closer to physical and technological realities. There used to be an old medical-school adage that in humans (or organisms generally) the physiology reflects the anatomy most at the center, that is, in the central nervous system. Something like it may also be granted to hold for computers: The function and the conditions of functioning reflect the structure most in the central processors.

Our extended hypothesis for language-as-bypass presupposes three levels, shown schematically in Figure 1; the L in L-level and sub L-level refers to language:

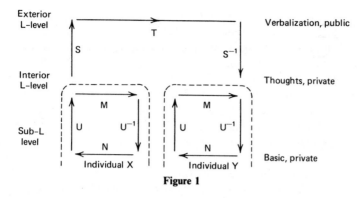

Figure 1

Whenever several levels exist, or are assumed to exist, the possibility of all corresponding bypasses ought to be considered. There are now two communication bypasses for the two individuals, the "shallow" one STS^{-1} and the "deep" one $USTS^{-1}U^{-1}$, as well as two partial-bypass possibilities: $STS^{-1}U^{-1}$ and $USTS^{-1}$. What, if anything, is represented by the purely internal bypasses UMU^{-1} and $U^{-1}NU$?

We have used earlier an ad hoc model concerning the late and early acquisition of language by children. This ad hoc model, which is really only an analogy and a very partial one at that, was outlined in terms of computer terminology. However, we ourselves are not computers or rather, we are more than computers. In particular, we have appetites and needs, we experience pleasure (at their satisfaction) and pain, besides being, like computers, processors of information. It might be hypothesized that at least some of such matters of the body are represented at what we call "basic level." If this is so then the basic level would acquire some features of what is traditionally referred to as body while the interior L-level might be associated in part with that other partner, that is, mind. Of course, there is really no dualism since no radical separation in anything like the Cartesian sense need be assumed.

It is tempting to consider the interpretation of the interior L-level as the "conscious" level since, presumably, whatever is in it can be verbalized and communicated to others. On the other hand, the interior basic level, being sub-L, would then receive the label of "unconscious." But what do we mean by "conscious," when we say that the interior L-level is "conscious"? It is rather that the two levels together, perhaps by means of something like the internal bypass UMU^{-1}, produce the novelty of "consciousness."

One aspect of it is its reflex character: If I am conscious of something I not only know it but I also know that I know it, and I know that too, and.... There seems to be hovering about consciousness something of the element of infinite regress. All of this seems to be taking us far from language and meaning. However, already in the introductory Chapter 1 we have said "Presumably, an adequate theory of meaning must be a court of last appeal: There is little use in trying to explain meaning if the explanation requires an explanation." We have already seen in Chapter 7, concerning Gödel's incompleteness theorem, how with the two levels, the metamathematical and the mathematical one, and with the round trip bypass between them, there arises one type of self-referentiality. By way of a farfetched analogy, the mathematical level there might be associated with the interior basic level here, and the metamathematical level with the interior L-level. This may be how the self-referentiality of "I am" arises. An interesting consequence would then be that consciousness is something relatively new: It cannot precede language. To sum up, the question What is language? has landed us in the question What is man? And our two-level model hints at an answer of this type: one who is two that are one.

Chapter Twelve

Summary and Conclusions

This twelfth and last chapter begins with a brief summary of the development of the main features of bypass. Convenient preliminary examples and illustrations were used in Chapter 1 to sketch what bypasses are and to provide their basic terminology. That initial sketch concentrated on the important notion of levels, on the three fundamental parts of a bypass, and on the three basic operations. The latter are the three ways of getting new bypasses out of old ones: by inverting (or switchover), by extending in length, and by stacking (which is extending in depth). They enable us to break down complex bypasses into simple parts and, conversely, to use simple bypasses in order to synthesize complex ones. Thus the three basic compositions supply a reversible-action glue for putting bypasses together and taking them apart or, to put it differently, they articulate bypasses. This alone goes some way toward justifying the claim made in the subtitle of this book that bypasses provide a simple approach to complexity.

Next to be mentioned are four important general types of bypasses. The first three were explicitly coded by the names i/c, SL, and CDC; the fourth one is hereby labeled I (which stands for involution or involutory). The i/c is a mnemonic for individual-collective, so that there should really be two subtypes: i/c/i and c/i/c. The idea is that relations between two individuals may (have to) be mediated through the collection(s) to which the individuals belong. Conversely, the relations between two collections may (have to) be arranged through some two representative individuals. Applications range from purely mathematical ones, such as assigning ordinal numbers to well-

225

ordered sets, to purely anthropological or political ones such as resolving an intertribal conflict by a fight between two champions or a negotiation between two representatives. It is remarked parenthetically that some remote connection to the i/c bypass may be traced in the ancient grammatical category of dual number, applying to nouns, verbs, and adjectives. This, in addition to singular and plural, was present in Sanskrit and old Greek; some of its traces remain in modern Polish and Lithuanian.

The second type is the SL-bypass where SL is a mnemonic for square-linear. The idea in applying this bypass is to escape the sheer numerical complexity of squared magnitudes N^2, where N is itself large, by a reduction to manageable linear magnitudes. So, with N individual telephone subscribers there are $N(N-1)/2$ possible pair connections but only N individual-to-central-exchange connections. The SL bypass appears as trunking and, through bypass stacking, as hierarchical trunking, in almost any large communication net. A closely related notion of collecting and distributing networks is described in Ex. 8.2. Something at least remotely resembling the SL-bypass seems to justify one of the conveniences of using general concepts; we only have to compare the statement "I saw a man on a horse" with "I saw John (or Bill, or Ivan, or your brother, or ...) on the young palomino (or the gray mare, or the old chestnut, or ...)." This suggests that there might be some theoretically and even practically interesting relations between the SL and the i/c types. To emphasize this possibility, one may recall here the approach to pattern recognition from Ex. 7.3.

The third general type to be noted is the CDC bypass, where CDC stands for the passage: continuous–discrete–continuous. The bypass switchover leads to the discrete–continuous–discrete round trip so that here, as with i/c, there should also be two subtypes: CDC and DCD. On the purely mathematical side, the DCD bypass is present in the method of generating functions whose very essence is the passage from a discrete sequence of coefficients to the continuous generating function, working with this function so as to determine it from a suitable functional equation, then disjoining it back into the discrete sequence of coefficients. The CDC type occurs mathematically when a suitably well-behaved continuous function is represented, say, as a Fourier series, then one works on the sequence of Fourier coefficients, and finally the coefficients are used to represent again a continuous function. The presence of CDC bypass in telecommunications is obvious: A continuous speech pattern is produced and received as continuous acoustic pattern. Yet the transmission is often done by discretizing voice into pulses, sending them, and then synthesizing these at the receiving end back into voice. Other examples of CDC bypass, in film and television, and in mediating the external events through the nervous system to our subjective impressions, are mentioned in Chapter 10. A special reference is made to the speculations, also from Chapter 10, on the possible role of the CDC bypass in the process of artistic creation.

An extensive collection is formed by those bypasses STS^{-1} in which the level-to-level transformation S is its own inverse: $S = S^{-1}$. A simple consequence is that S repeated twice in succession yields the identity: $S^2 = 1$. Since transformations with that property are called in mathematics involutions, our collection might be called involution bypasses, or I-bypasses for short. There are various examples: Some mathematical ones are given in Chapters 4–6; the principle of reductio ad absurdum is treated from this point of view in Chapter 2; this treatment is followed in Chapter 2 by a suggestion that the same sort of thing is involved in the use of seemingly abusive language between people on intimate basis, and also in satire and as a means for avoiding hubris. There are examples in technology, for instance, with switching devices. More generally, any two-state device is an example: If S stands for the shift from one state to the other, then S repeated twice brings back the original state: $S^2 = 1$. There is a class of codes in which the same transformation that enciphers the message will also decipher it on being applied to the encoded text. A simple instance is obtained by transposing the first two letters, then the next two, and so on: htsu ew boatni rtnapssotioisn. In the modest and important domestic domain, some examples of sewing and mending are I-bypasses: Turn the sock or shirt inside out—do the mending or sewing—turn it inside out again. It is not surprising that this procedure is also useful in some industrial manufacturing processes. What may be mildly surprising is that almost the same double turning inside out is sometimes mathematically useful. A tiny example is in obtaining the left-open right-closed staircase function $-[1 - x]$ from the usual staircase function $[x]$ (that is, the greatest integer $\leqslant x$), which is left-closed and right-open.

There seems to be a huge gap between mending socks and Jacobi's principle "one must always invert" as applied to the theory of elliptic integrals and functions. If, to quote E. M. Forster, one must "only connect," then the bypass principle achieves this, for it connects after a fashion sock mending and Jacobi's inversion. So, some properties of elliptic integrals may be conveniently obtained by first inverting them to get elliptic functions, working with these, and then inverting again to return to the integrals. Together, sock mending and Jacobi's inversion suggest a generally promising tactic: to investigate transformations S that invert, say to S^{-1}, and to interpolate some useful middle part T between S and S^{-1}. This, of course, leads to the bypass STS^{-1}.

The four types just discussed (i/c, SL, CDC, I) suggest the beginnings of certain methods of ordering and grouping bypasses, for instance, by context, by levels, or by the form of the outer part. Of course, an important basis for classification might be by purpose or function. There are various candidates here and we mention three (not necessarily exclusive ones): equivalence, transport, reflexness. With the equivalence bypasses, the purpose is to establish that two objects (or descriptions, or views, or transformations) that appear to be different refer to one and the same thing, though perhaps in two different ways or frameworks. The canonical mathematical example is algebraic: the similarity of matrices in Ex. 6.1, which is perhaps the purest of all bypasses. We recall that it settles the fundamental equivalence question: When is it true that two matrices describe one and the same linear transformation of a vector space, written down relative to some two of its bases? There are also two canonical physical examples, related to the preceding algebraic one: The first one concerns inertial frames in special relativity and the Lorentz transformation (Ex. 9.5); the other one asserts the equivalence of the Schrödinger and the Heisenberg pictures in quantum mechanics (Ex. 9.6).

On general, rather than specialized mathematical or physical, grounds the equivalence bypass might serve as a sort of a paradigm for reconciliation. To see this let us consider, say, two seemingly different or even opposing views X and Y. Suppose that with our superior wisdom and objectivity we see that the differences between X and Y are only incidental and due to some details which do not really matter. How could reconciliation be brought about and understanding achieved to the satisfaction of the two parties holding the views? A valuable hint is provided by such ordinary expressions as "common ground," "talking the same language," and even "being on the same wavelength." A further hint, less ordinary but also supportive, comes from the *characteristica universalis* of Leibniz. One recalls his motto *"calculemus!"* (let us calculate), which Leibniz regarded as indicating a possibility of resolving national conflicts by means of his (projected) universal logic. It may happen that we, from our vantage point, can provide a passage S from the level at which the views X and Y are expressed and seem to differ, to our own level, together with the reverse passage S^{-1}. It may further happen that the two disagreeing parties are able to follow the passages S and S^{-1} provided for them. Then, in effect, the way is open to reconciling X and Y by an equivalence bypass through our own level.

Perhaps the most important, and certainly the commonest, reason for bypasses is one form or another of transport. Already the preliminary descriptions and pseudo-definitions of Chapter 1 rely on representing bypass as transport past an obstacle or difficulty. Many of our physical and technological examples concern transport of objects, matter, energy, or information. In the mathematical examples it is often structure that is transported, and some of the more important ones are important precisely because of new structure or new ways of introducing structure. The few very crude examples given from economics are about transport of money, value, or credit. The group consisting of metaphors, similes, analogies, allegories, kennings, and parables was considered on Chapter 11. In this connection, as also with some other topics of Chapter 11, one may hesitate, with the angels, to assert that all this concerns transport of meaning and structure, but one need not be a fool in believing that a strong element of transporting both is undeniably present. Finally, there are some strange types of transport bypasses; the examples from humor and from theology, given in Chapter 1, come to mind here. It might be granted that the former seem sometimes to concern a transport of nothing, and the latter a transport of grace.

We finish the subject of strange transports with a very doubtful instance from politics. Suppose that a group of people elects representatives to the governing body that deliberates on important matters and settles them, passing its decisions as generally binding. Is this a case of transport of political power through a bypass STS^{-1} in which S stands for "upward" delegation of power, T for the deliberating and deciding, and S^{-1} for "downward" exercise of power? Some aspects of the bypass principle, such as stacking (with reference to intermediate administrative or power levels) and possible connections with the SL and i/c types would seem to suggest an affirmative answer to the question. However, it is not at all clear whether we have here an example of a bypass or a counterexample of something that seems to be one but is not. In particular, it is a moot point whether or not the above group of people might be said to be self-governing. This brings us to the subject of reflexness.

If transport is the commonest purpose of bypasses, reflexness is perhaps the subtlest. This lies in the very nature of reflex, or self-referential or boot-strapping action: The agent or subject acts, or appears to act, not on others but on itself. It moves itself, refers to itself, copies itself, or relates itself to itself. The inevitable price to pay in passing from simple action on others to complex action on oneself is, of course, complexity. Since the bypass principle is essentially an approach to complexity it should perhaps not be surprising that it is concerned in reflexness. Let us start with the trite observation that one cannot lift oneself by pulling on one's own bootstraps. In the physical domain this is forbidden by Newton's third law and highlighted further by the Archimedean dictum: Give me whereon to stand and I shall move the earth. In logic self-reference is outlawed to prevent the occurrence of para-doxes of various types. In matters of life and birth, whatever answer is given to the pseudo-problem of the chicken and the egg, living things produce replicas of themselves, but not themselves. Even in the psycho-literary domain it may be that certain types of incest, such as the case of Oedipus, are particularly offensive because they interfere with the orderly progression of generations; thus Oedipus is cast in the impossible role of being his own ancestor. In the sociology of labor and education (with reference to Hawthorne-type experiments) the workers do not lift themselves into a higher productive gear, or students into better understanding, by themselves.

To continue with the cliché on lifting oneself by one's own bootstraps, only simple or naive types of bootstrapping are impossible or forbidden. To put it roughly, an isolated system or a system functioning at one level only cannot be reflex. But if the system operates on two levels or if it has suitable access to a level outside its own, the matters may be different, and reflexness of some sort might occur. Even the literal bootstrapping is possible if another level is used: We can lift ourselves off the floor if we are allowed to attach something to the ceiling. Perhaps even the "double" in the "double helix" structure of the self-replicating DNA molecule has some remote connection with our "two levels." The Hawthorne effect in the social domain might be viewed as a reflex bypass whereby the experimental subjects (workers or students) raise themselves, temporarily perhaps, into a new state. This favors higher output or comprehension and is achieved not so much by the methods applied in the experiment as by a complex interaction with the experimenters who may stand for "the establishment." Ultimately, perhaps, the Hawthorne effect acts here as a powerful technique to self-combat boredom. The obvious and flattering attention of the experimenters to the subjects induces the reciprocal attention in the subjects, hence grounds for reflexness. If the above interpretation of the Hawthorne effect is anywhere near being correct, then it may be not too much of an exaggeration to say that the effort to apply the effect, make it widespread, and possibly lasting, should be one of the principal social aims. It is as though the religious injunction "Love ye one another" were reinterpreted as a logical argument for altruism: One cannot really love or admire oneself in any effective sense but only reflexly, through others.

Our first example of reflexness, the rawest and merely superficially expository, was the tank example in Chapter 1. The doubtful claim to reflexness may perhaps be justified thus: The tank (one level) moves on the bridge (another level) of its cleats. But the chain of the cleats closes up so that in effect the tank functions by carrying its own "other level" along with itself. The next example was mathematical and concerned the use of generating functions from Ex. 7.1, where relexness appears as self-relating achieved through DCD bypass. At the D-level (D for discrete) there is a recursion, that is, a relation of objects to some other, preceding, objects. In more detail, the first k coefficients of an infinite sequence are known, and thereafter each one is given in terms of the previous k ones. At the C-level (C for collective or continuous), the continuous generating function of the sequence is related to itself. One of the best examples of reflexness is the apparently self-referring proposition from Gödel's first incompleteness theorem (Ex. 7.2). It was claimed on its behalf, and the claim was supported by quotations from Gödel's own work, that the self-referentiality arises here too by a certain round trip passage of the interlevel type. It starts from the metamathematical level (of theorems about theorems) and proceeds to the mathematical level (of theorems about numbers), diagonalization is performed then, and finally comes the return to the metamathematical level. The crucial device of Gödel numbering is used twice, once directly for encoding and once in reverse for decoding; it provides a language of sorts, for interlevel communication. This brings to mind language itself, that is, natural language like English (or French or Chinese), for here we have the commonest case of reflexness: It is possible to talk in English about English. If language is indeed a two-level affair, as was claimed in the last chapter, then the possibility for reflexness may come from the nature of language itself rather than be dependent on its highly refined development and subtle use.

At the end of the last chapter, a claim was made that the question What is language? leads to the ancient question What is man? This was followed by a highly tentative hint concerning the two-levels mode of human function. There are many reformulations of that ancient question, and we propose now one more: What is man that he should have a partial access to infinity? Our reformulation comprehends the important matter of novelty and innovative power, that is, of infinite human productivity. The question What is man? is inescapably paired with the question of human evolution: How did man develop? Here a safe answer, though one that does not by itself explain much, is probably the following: Man evolved by learning to cope with complexity. A detailed explanation or even an adequate outline, of such learning is, of course, an enormous task. The conjugacy principle seems to offer a modeling hope here in several different respects. The first one concerns our temporary holding devices (fifth preliminary example in Chapter 1 and Ex. 8.4) and especially that most universal one of such devices, which is the human hand. The second one relates to transport and containers (Ex. 8.2). The third one deals with some foundations of mental activity of the logical type (patterns, from the end of Ex. 7.3 and maps from Ex. 8.3). The fourth one refers to language (Chapter 11). It may even be that the bypass principle, in its various aspects, was so successful a means of coping with complexity that the evolving hominid ended up by interiorizing it, and so became man.

To finish our lengthy chain of speculations, we return now to our reformulation: What is man that he should have a partial access to infinity? It is characteristic of many reflexness bypasses that they involve or produce an infinitude somewhere, a sort of infinite closure. As we would claim, this is because of the two-levels mode of operation. Perhaps the simplest example is the infinity of reflections produced by two parallel mirrors facing each other. A very clear case of infinite closure occurs in the example of generating functions, and a technically more complicated case is with the diagonalization of the Gödel theorem. Even our primitive example of the tank involves a sort of infinity: The chain of its cleats closes up in a circle. So here we have a basis, however tenuous, for the partial access to infinity and for the reflex nature of our consciousness. They may come from the interiorizing suggested earlier and from the two interior levels suggested at the end of Chapter 11.

We have tried to convey the promise and the excitement that go with the bypass principle and make it a subject worth developing for its practical use as well as for its theoretical interest. Both would plainly call for much critical further work from many points of view. It is to promote general, specialist, and interdisciplinary interest toward that end that this book was written.

References

[1] Abel, N. H. *Oeuvres Complètes*, 2nd ed., Christiania, 1881.

[2] Ackermann, W. *Math. Annal.*, **99**: 118, (1929).

[3] André, D. *Comptes Rendus*, **88**: 965, (1879).

[4] Anttila, R. *An Introduction to Historical and Comparative Linguistics*, Macmillan, New York, 1972.

[5] Armstrong, E. H. *Proc. I.R.E.*, **24**: 689, (1936).

[6] Artin, E. *Geometric Algebra*, Interscience, New York, 1951.

[7] Artin, E. *The Gamma Function*, Holt, Rinehart, and Winston, New York, 1964.

[8] Auslander, L. and Mackenzie, R. E. *Introduction to Differentiable Manifolds*, McGraw-Hill, New York, 1963.

[9] Babbage, C. See Ch. 4 in [44].

[10] Babbage, C. *Passages from the Life of a Philosopher*, Longman and Co., London, 1864.

[11] Banach, S. and Tarski, A. *Fund. Math.*, **6**: 244, (1924).

[12] Barlow, A. E. *The History and Principles of Weaving*, Sampson Low, Marston, Searle and Rivington, London, 1878.

[13] Barwise, J. (ed.) *Handbook of Mathematical Logic*, North-Holland, Amsterdam, 1978.

[14] Bateman, H. et al. *Higher Transcendental Functions*, McGraw-Hill, New York, 1953.

[15] Beth, E. W. *The Foundations of Mathematics*, North-Holland, Amsterdam, 1959.

[16] Birkhoff, G. and MacLane, S. *A Survey of Modern Algebra*, Macmillan, New York, 1948.

[17] Black, H. S. *Modulation Theory*, Van Nostrand, Princeton, N.J., 1953.

[18] Blaschke, W. *Kreis und Kugel*, Chelsea, New York, 1949.

[19] Blumenthal, L. and Menger, K. *Studies in Geometry*, W. H. Freeman, San Francisco, 1971.

[20] Boas, R. R. *Entire Functions*, Academic Press, New York, 1954.

[21] Bödewadt, U. T. *Math. Z.*, **49**: 497, (1943).

[22] Boothby, W. M. *An Introduction to Differentiable Manifolds*, Academic, New York, 1975.

[23] Bourge, R. and Azza, J. P. *Écrites et Mémoires Mathématiques d'Évariste Galois*, Gauthier-Villars, Paris, 1962.

[24] Boyer, C. B. *A History of Mathematics*, Wiley, New York, 1968.

[25] Buxton, H. W. *Buxton Mss.*, Ashmolean Museum for the History of Science, Oxford.

[26] Cantor, G. *Gesammelte Abhandlungen*, Georg Olms, 1962.

[27] Cantor, M. *Vorlesungen über die Geschichte der Mathematik*, Johnson Reprint Corp., 1965.

[28] Canby, E. T. *A History of Electricity*, Hawthorne Books, 1963.

[29] Capra, F. *The Tao of Physics*, Bantam Books, New York, 1977.

[30] Churchill, R. V. *Complex Variables and Applications*, 2nd ed., McGraw-Hill, New York, 1960.

[31] Coddington, E. and Levinson, N. *Theory of Ordinary Differential Equations*, McGraw-Hill, New York, 1955.

[32] Coulanges, N. D. F. de, *The Ancient City*, Doubleday, Garden City, N.Y., 1956.

[33] Courant, R. and Hilbert, D. *Methoden Der Mathematischen Physik*, 2nd ed., Springer, Berlin, 1931.

[34] Crowley, T. H. et al. *Modern Communications*, Columbia University Press, New York, 1962.

[35] Cusanus, N. *De Visione Dei*, Strassbourg 1488, Atlantic Paperbacks, 1960.

[36] Dascal, M. *Studia Leibnitiana*, Steiner, 1976, vol. 8, 187.

[37] Davis, M. *Computability and Unsolvability*, McGraw-Hill, New York, 1958.

[38] Dienes, P. *The Taylor Series*, Dover, New York, 1957.

[39] Dilworth, R. P. *Ann. Math.*, **57**: 161, (1950).

[40] Disraeli, B. *Coningsby*, London, 1844.

[41] Doetsch, G. *Theorie und Anwendung der Laplace Transformation*, Dover, New York, 1943.

[42] Dörrie, H. *Hundred Great Problems*, 2nd ed., Dover, New York, 1963.

[43] Doyon, A. and Liaigre, L. *Jacques Vaucanson*, Presses Universitaires de France, 1966.

[44] Dubbey, J. M. *The Mathematical Work of Charles Babbage*, Cambridge University Press, Cambridge, England, 1978.

[45] *Encylopedia Britannica*, 11th ed.

[46] Eves, H. *A Survey of Geometry* (rev. ed.), Allyn and Bacon, Boston, 1972.

[47] Ewing, J. A. *Philos. Trans. Soc. London*, **176**: 523, (1886).

[48] Fagen, M. D. (ed.) *A History of Engineering and Science in the Bell System*, Bell Laboratories, 1975, Vol. 1.

[49] Feller, W. *Introduction to Probability Theory and Its Applications*, 2nd ed., Wiley, New York, 1957.

[50] Feynman, R. P. et al. *The Feynman Lectures On Physics*, Addison-Wesley, Reading, Mass., 1965.

[51] Frisch, H. L. and Hammersley, J. M. *SIAMJ Appl. Math.*, **11**: 894, (1963).

[52] Gauss, C. F. *Disquisitiones Arithmeticae* (Eng. trans.), Yale University Press, New Haven, Conn., 1966.

[53] Gennep, A. van, *The Rites of Passage*, University of Chicago Press, Chicago, 1960.

[54] Gilman, R. H. *Sci. Am.*, p. 81, October 1968.

[55] Gleason, A. *Bull. A.M.S.*, **55**: 446, (1949).

[56] Gödel, K., *The Consistency of the Continuum Hypothesis*, Princeton University Press, Princeton, N.J., 1940.

[57] Gödel, K. *Monatsch. Math. Phys.*, **38**: 173, (1931).

[58] Goldstein, H. *Classical Mechanics*, Addison-Wesley, Reading, Mass., 1950.

[59] Gottfried, K. *Quantum Mechanics*, W. A. Benjamin, New York, 1974, Vol. 1.

[60] Goursat, E. *Mathematical Analysis*, Ginn, Boston, 1904, Vol. 2, part 2.

[61] Hall, M. *The Theory of Groups*, Macmillan, New York, 1959.

[62] Halmos, P. *Measure Theory*, Van Nostrand, Princeton, N.J., 1950.

[63] Hardy, G. H. and Wright, E. M. *An Introduction to The Theory of Numbers*, 2nd ed., Clarendon, Oxford, 1945.

[64] Hausdorff, F. *Math. Annal.*, **94**: 244, (1925).

[65] Hausdorff, F. *Mengenlehre* 3rd ed., Dover, New York.

[66] Heck, C. *Magnetic Materials and Their Applications*, Butterworths, London, 1974.

[67] Hermes, H. *Enumerability, Decidability, Computability*, Springer, Berlin, 1965.

[68] Herstein, I. N. *Topics in Algebra*, Blaisdell, Waltham, 1964.

[69] Hilbert, D. *Grundlagen der Geometrie*, 8th ed., Stuttgart, 1956.

[70] Hilbert, D. *Math. Annal.*, **95**: 161, (1926).

[71] Hille, E. and Phillips, R. S. *Functional Analysis and Semi-Groups*, American Mathematical Society Providence, R.I., 1957.

[72] Hölder, O. *Math. Annal.*, **28**: 1, (1887).

[73] Humphries, S. *Some Persian Carpets*, A. and C. Black, 1910.

[74] International Mathematical Congress Acts, Nice, 1970.

[75] Isaac, G. *Sci. Am.*: 90 (1978).

[76] Jacobi, C. G. J. *Neue Methode zur Integration ...*, Ostwald Klassiker, Engelmann, 1906.

[77] Jevons, W. S., *The Principles of Science*, Macmillan, London, 1870.

[78] Kahn, D. *The Codebreakers*, Weidenfeld and Nicolson, 1967.

[79] Kirk, G. S. and Raven, J. E. *The Presocratic Philosophers*, Cambridge University Press, Cambridge, England, 1957.

[80] Kleene, S. C. *Introduction to Metamathematics*, Van Nostrand, Princeton, N. J., 1956.

[81] Knowlton, K. and Harmon, L. *Computer Graphics and Image Processing*, 1972, Vol. 1, p. 1.

[82] Korzybski, A. *Science and Sanity*, 4th ed., The International Non-Artistotelian Library, 1958.

[83] Kuhn, H. W. and Kuenne, R. E. *Regional Sci.*, **4**: 21, (1962).

[84] Kurosh, A. G. *The Theory of Groups*, 2nd ed., Chelsea, New York, 1960.

[85] Landau, L. and Rümer, E. See References in [130].

[86] Lewin, K. *Principles of Topological Psychology*, McGraw-Hill, New York, 1936.

[87] Lewin, K. *Field Theory in Social Science*, Harper, New York, 1951.

[88] Lighthill, J. et al. (eds) *Telecommunications in the 1980's and After*, The Royal Society, London, 1978.

[89] Littlewood, J. E. *A Mathematician's Miscellany*, Methuen, London, 1953.

[90] Lloyd, G. E. R. *Polarity and Analogy*, Cambridge University Press, Cambridge, England, 1966.

[91]   Lyndon, R. C. and Shupp, P. E. *Combinatorial Group Theory*, Springer, Berlin, 1977.

[92]   MacLane, S. *Categories for the Working Mathematician*, Springer, Berlin, 1971.

[93]   Marrou, H. I. *A History of Education in Antiquity*, Mentor, 1964.

[94]   Mason, J. A. *The Ancient Civilizations of Peru*, Penguin, England, 1957.

[95]   Melzak, Z. A. *Companion to Concrete Mathematics*, Wiley, New York, 1973.

[96]   Melzak, Z. A. *Mathematical Ideas, Modeling, and Applications*, Wiley, New York, 1976.

[97]   Melzak, Z. A. and Forsyth, J. S. *Q. Appl. Math.*, **35**: 239, (1977).

[98]   Melzak, Z. A. *Proc. Cambridge Philos. Soc.*, **86**: 313, (1979).

[99]   Mikusinski, J. *Operational Calculus*, Pergamon, New York, 1959.

[100] Miller, G. et al. *Plans and the Structure of Behavior*, Holt, New York, 1960.

[101] Montgomery, D. and Zippin, L. *Topological Transformation Groups*, Interscience, New York, 1955.

[102] Montgomery, D. and Zippin, L. *Ann. Math.*, **56**: 213, (1952).

[103] Moon, J. W. in *Seminar on Graph Theory*, edited by F. Harary, Holt, Rinehart and Winston, New York, 1967.

[104] Moseley, M. *Irascible Genius*, Hutchinson, London, 1964.

[105] Neumann, J. von *Theory of Self-Reproducing Automata*, University of Illinois Press, Urbana, Ill., 1966.

[106] Palmer, L. R. *Descriptive and Comparative Linguistics*, Faber and Faber, London, 1978.

[107] Peano, G. *Arithmetices Principia*, Bocca, Turin, 1889.

[108] Pei, M. *Glossary of Linguistic Terminology*, Columbia University Press, New York, 1966.

[109] Perkus, J. K. *Combinatorial Methods*, Springer, Berlin, 1971.

[110] Pierce, J. R. and Karlin, J. E. *The Bell Syst. Tech. J.*, p. 497, 1957.

[111] Polya, G. and Szegö, G., *Problems and Theorems in Analysis*, Springer, Berlin, 1972, Vol. 1.

[112] Porta, G. B. della *Natural Magick*, London 1658 (facsimile ed.) Basic Books, New York, 1957.

[113] Preminger, A. (ed.) *Princeton Encyclopedia of Poetry and Poetics*, Princeton University Press, Princeton, N.J., 1974.

[114] Priestley, J. *A Course of Lectures on Oratory and Criticism*, (photocopy ed.) Garland, 1971.

[115] Prinz, E. *Bull. Schweiz. Elektrototechn. Ver.*, **65**: 1, (1975).

[116] Pritchard, J. B. (ed.) *Ancient Near Eastern Texts*, 2nd ed., Princeton University Press, Princeton, N.J., 1955.

[117] Raby, F. J. E. (ed.) *The Oxford Book of Medieval Latin Verse*, Oxford University Press, 1959.

[118] Ramsey, F. P. *The Foundations of Mathematics*, Harcourt, Brace & World, New York, 1931.

[119] Riley, H. T. *Dictionary of Latin Quotations*, Bohn, 1860.

[120] Riordan, J. *An Introduction to Combinatorial Analysis*, Wiley, New York, 1958.

[121] Ritt, J. F. *Integration in Finite Terms*, Columbia University Press, New York, 1948.

[122] Ritt, J. F. *Am. J. Math.*, **21**: 348, (1920).

[123] Rota, G. C. *J. Comb. Theor. Ser. A*, **24**: 395, (1978).

[124] Routh, E. J. *A Treatise on Analytical Statics*, 2nd ed., Cambridge University Press, Cambridge, England, 1896, Vol. 1.

[125] Rund, H. *The Hamilton-Jacobi Theory in The Calculus of Variations*, Huntington, 1973.

[126] Schröder, E. *Math. Annal.*, **3**: 296, (1871).

[127] Shannon, C. E. *The Bell Syst. Tech. J.*, July: 397 (1948).

[128] Siegel, C. L. *Topics in Complex Variable Theory*, Wiley, New York, 1969, Vol. 1.

[129] Skelton, R. A. et al. (eds.) *The Vinland Map and the Tartar Relation*, Yale University Press, New Haven, Conn., 1965.

[130] Sneddon, I. *Fourier Transforms*, McGraw-Hill, New York, 1951.

[131] Stam, J. H. *Inquiries into the Origin of Language*, Harper & Row, New York, 1976.

[132] Still, A. *Communication Through the Ages*, Murray Hill Books, 1946.

[133] Talley, D. *Basic Telephone Switching Systems*, Hayden, 1979.

[134] Titchmarsh, E. C. *The Theory of Functions*, 2nd ed., Oxford University Press, 1939.

[135] Turan, P., personal communication.

[136] Ulam, S. M. *Adventures of a Mathematician*, Charles Scribner's Sons, New York, 1976.

[137] Uspensky, J. V. and Heaslet, M. *Elementary Number Theory*, McGraw-Hill, New York, 1939.

[138] Vickers, B. *Classical Rhetoric in English Poetry*, Macmillan, London, 1970.

[139] Weber, A. *Theory of the Location of Industries*, Chicago, 1929.

[140] Weedy, B. M. *Electric Power Systems*, 2nd ed., Wiley, New York, 1970.

[141] Wilkins, J. *Mercury; or, Secret and Swift Messenger* (facsimile ed.), Frank Cass, London, 1970.

[142] Zassenhaus, H. J. *The Theory of Groups*, Chelsea, New York, 1949.

Index